Peter Dalgaard

Introductory Statistics with R

Second Edition

Peter Dalgaard
Department of Biostatistics
University of Copenhagen
Denmark
p.dalgaard@biostat.ku.dk

ISSN: 1431-8784
ISBN: 978-0-387-79053-4   e-ISBN: 978-0-387-79054-1
DOI: 10.1007/978-0-387-79054-1

Library of Congress Control Number: 2008932040

Printed on acid-free paper

springer.com

Statistics and Computing

Series Editors:
J. Chambers
D. Hand
W. Härdle

Statistics and Computing

To Grete, for putting up with me for so long

Preface

R is a statistical computer program made available through the Internet under the General Public License (GPL). That is, it is supplied with a license that allows you to use it freely, distribute it, or even sell it, as long as the receiver has the same rights and the source code is freely available. It exists for Microsoft Windows XP or later, for a variety of Unix and Linux platforms, and for Apple Macintosh OS X.

R provides an environment in which you can perform statistical analysis and produce graphics. It is actually a complete programming language, although that is only marginally described in this book. Here we content ourselves with learning the elementary concepts and seeing a number of cookbook examples.

R is designed in such a way that it is always possible to do further computations on the results of a statistical procedure. Furthermore, the design for graphical presentation of data allows both no-nonsense methods, for example `plot(x,y)`, and the possibility of fine-grained control of the output's appearance. The fact that R is based on a formal computer language gives it tremendous flexibility. Other systems present simpler interfaces in terms of menus and forms, but often the apparent user-friendliness turns into a hindrance in the longer run. Although elementary statistics is often presented as a collection of fixed procedures, analysis of moderately complex data requires ad hoc statistical model building, which makes the added flexibility of R highly desirable.

R owes its name to typical Internet humour. You may be familiar with the programming language C (whose name is a story in itself). Inspired by this, Becker and Chambers chose in the early 1980s to call their newly developed statistical programming language S. This language was further developed into the commercial product S-PLUS, which by the end of the decade was in widespread use among statisticians of all kinds. Ross Ihaka and Robert Gentleman from the University of Auckland, New Zealand, chose to write a reduced version of S for teaching purposes, and what was more natural than choosing the immediately preceding letter? Ross' and Robert's initials may also have played a role.

In 1995, Martin Maechler persuaded Ross and Robert to release the source code for R under the GPL. This coincided with the upsurge in Open Source software spurred by the Linux system. R soon turned out to fill a gap for people like me who intended to use Linux for statistical computing but had no statistical package available at the time. A mailing list was set up for the communication of bug reports and discussions of the development of R.

In August 1997, I was invited to join an extended international core team whose members collaborate via the Internet and that has controlled the development of R since then. The core team was subsequently expanded several times and currently includes 19 members. On February 29, 2000, version 1.0.0 was released. As of this writing, the current version is 2.6.2.

This book was originally based upon a set of notes developed for the course in Basic Statistics for Health Researchers at the Faculty of Health Sciences of the University of Copenhagen. The course had a primary target of students for the Ph.D. degree in medicine. However, the material has been substantially revised, and I hope that it will be useful for a larger audience, although some biostatistical bias remains, particularly in the choice of examples.

In later years, the course in Statistical Practice in Epidemiology, which has been held yearly in Tartu, Estonia, has been a major source of inspiration and experience in introducing young statisticians and epidemiologists to R.

This book is not a manual for R. The idea is to introduce a number of basic concepts and techniques that should allow the reader to get started with practical statistics.

In terms of the practical methods, the book covers a reasonable curriculum for first-year students of theoretical statistics as well as for engineering students. These groups will eventually need to go further and study more complex models as well as general techniques involving actual programming in the R language.

For fields where elementary statistics is taught mainly as a tool, the book goes somewhat further than what is commonly taught at the under-graduate level. Multiple regression methods or analysis of multifactorial experiments are rarely taught at that level but may quickly become essential for practical research. I have collected the simpler methods near the beginning to make the book readable also at the elementary level. However, in order to keep technical material together, Chapters 1 and 2 do include material that some readers will want to skip.

The book is thus intended to be useful for several groups, but I will not pretend that it can stand alone for any of them. I have included brief theoretical sections in connection with the various methods, but more than as teaching material, these should serve as reminders or perhaps as appetizers for readers who are new to the world of statistics.

Notes on the 2nd edition

The original first chapter was expanded and broken into two chapters, and a chapter on more advanced data handling tasks was inserted after the coverage of simpler statistical methods. There are also two new chapters on statistical methodology, covering Poisson regression and nonlinear curve fitting, and a few items have been added to the section on descriptive statistics. The original methodological chapters have been quite minimally revised, mainly to ensure that the text matches the actual output of the current version of R. The exercises have been revised, and solution sketches now appear in Appendix D.

Acknowledgements

Obviously, this book would not have been possible without the efforts of my friends and colleagues on the R Core Team, the authors of contributed packages, and many of the correspondents of the e-mail discussion lists.

I am deeply grateful for the support of my colleagues and co-teachers Lene Theil Skovgaard, Bendix Carstensen, Birthe Lykke Thomsen, Helle Rootzen, Claus Ekstrøm, Thomas Scheike, and from the Tartu course Krista Fischer, Esa Läära, Martyn Plummer, Mark Myatt, and Michael Hills, as well as the feedback from several students. In addition, several people, including Bill Venables, Brian Ripley, and David James, gave valuable advice on early drafts of the book.

Finally, profound thanks are due to the free software community at large. The R project would not have been possible without their effort. For the typesetting of this book, T_EX, L^AT_EX, and the consolidating efforts of the L^AT_EX2e project have been indispensable.

Peter Dalgaard
Copenhagen
April 2008

Contents

1

Basics

The purpose of this chapter is to get you started using R. It is assumed that
you have a working installation of the software and of the ISwR package
that contains the data sets for this book. Instructions for obtaining and
installing the software are given in Appendix A.

The text that follows describes R version 2.6.2. As of this writing, that is
the latest version of R. As far as possible, I present the issues in a way
that is independent of the operating system in use and assume that the
reader has the elementary operational knowledge to select from menus,
move windows around, etc. I do, however, make exceptions where I am
aware of specific difficulties with a particular platform or specific features
of it.

1.1   First steps

This section gives an introduction to the R computing environment and
walks you through its most basic features.

Starting R is straightforward, but the method will depend on your computing platform. You will be able to launch it from a system menu, by
double-clicking an icon, or by entering the command "R" at the system
command line. This will either produce a console window or cause R
to start up as an interactive program in the current terminal window. In

P. Dalgaard, *Introductory Statistics with R*,
DOI: 10.1007/978-0-387-79054-1_1, © Springer Science+Business Media, LLC 2008

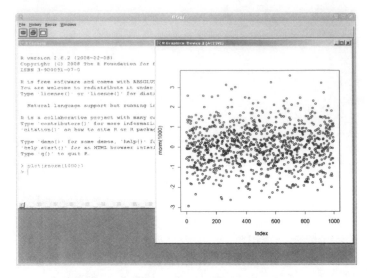

Figure 1.1. Screen image of R for Windows.

either case, R works fundamentally by the question-and-answer model: You enter a line with a command and press Enter (↩). Then the program does something, prints the result if relevant, and asks for more input. When R is ready for input, it prints out its prompt, a ">". It is possible to use R as a text-only application, and also in batch mode, but for the purposes of this chapter, I assume that you are sitting at a graphical workstation.

All the examples in this book should run if you type them in exactly as printed, *provided* that you have the ISwR package not only installed but also loaded into your current search path. This is done by entering

```
> library(ISwR)
```

at the command prompt. You do not need to understand what the command does at this point. It is explained in Section 2.1.5.

For a first impression of what R can do, try typing the following:

```
> plot(rnorm(1000))
```

This command draws 1000 numbers at random from the normal distribution (rnorm = *r*andom *norm*al) and plots them in a pop-up graphics window. The result on a Windows machine can be seen in Figure 1.1.

Of course, you are not expected at this point to guess that you would obtain this result in that particular way. The example is chosen because it shows several components of the user interface in action. Before the style of commands will fall naturally, it is necessary to introduce some concepts and conventions through simpler examples.

Under Windows, the graphics window will have taken the keyboard focus at this point. Click on the console to make it accept further commands.

1.1.1 An overgrown calculator

One of the simplest possible tasks in R is to enter an arithmetic expression and receive a result. (The second line is the answer from the machine.)

```
> 2 + 2
[1] 4
```

So the machine knows that 2 plus 2 makes 4. Of course, it also knows how to do other standard calculations. For instance, here is how to compute e^{-2}:

```
> exp(-2)
[1] 0.1353353
```

The [1] in front of the result is part of R's way of printing numbers and vectors. It is not useful here, but it becomes so when the result is a longer vector. The number in brackets is the index of the first number on that line. Consider the case of generating 15 random numbers from a normal distribution:

```
> rnorm(15)
 [1] -0.18326112 -0.59753287 -0.67017905  0.16075723  1.28199575
 [6]  0.07976977  0.13683303  0.77155246  0.85986694 -1.01506772
[11] -0.49448567  0.52433026  1.07732656  1.09748097 -1.09318582
```

Here, for example, the [6] indicates that 0.07976977 is the sixth element in the vector. (For typographical reasons, the examples in this book are made with a shortened line width. If you try it on your own machine, you will see the values printed with six numbers per line rather than five. The numbers themselves will also be different since random number generation is involved.)

1.1.2 Assignments

Even on a calculator, you will quickly need some way to store intermediate results, so that you do not have to key them in over and over again. R, like other computer languages, has *symbolic variables*, that is names that can be used to represent values. To assign the value 2 to the variable x, you can enter

```
> x <- 2
```

The two characters <- should be read as a single symbol: an arrow pointing to the variable to which the value is assigned. This is known as the *assignment operator*. Spacing around operators is generally disregarded by R, but notice that adding a space in the middle of a <- changes the meaning to "less than" followed by "minus" (conversely, omitting the space when comparing a variable to a negative number has unexpected consequences!).

There is no immediately visible result, but from now on, x has the value 2 and can be used in subsequent arithmetic expressions.

```
> x
[1] 2
> x + x
[1] 4
```

Names of variables can be chosen quite freely in R. They can be built from letters, digits, and the period (dot) symbol. There is, however, the limitation that the name must not start with a digit or a period followed by a digit. Names that start with a period are special and should be avoided. A typical variable name could be height.1yr, which might be used to describe the height of a child at the age of 1 year. Names are case-sensitive: WT and wt do not refer to the same variable.

Some names are already used by the system. This can cause some confusion if you use them for other purposes. The worst cases are the single-letter names c, q, t, C, D, F, I, and T, but there are also diff, df, and pt, for example. Most of these are functions and do not usually cause trouble when used as variable names. However, F and T are the standard abbreviations for FALSE and TRUE and no longer work as such if you redefine them.

1.1.3 Vectorized arithmetic

You cannot do much statistics on single numbers! Rather, you will look at data from a group of patients, for example. One strength of R is that it can handle entire *data vectors* as single objects. A data vector is simply an array of numbers, and a vector variable can be constructed like this:

```
> weight <- c(60, 72, 57, 90, 95, 72)
> weight
[1] 60 72 57 90 95 72
```

The construct `c(...)` is used to define vectors. The numbers are made up but might represent the weights (in kg) of a group of normal men.

This is neither the only way to enter data vectors into R nor is it generally the preferred method, but short vectors are used for many other purposes, and the `c(...)` construct is used extensively. In Section 2.4, we discuss alternative techniques for reading data. For now, we stick to a single method.

You can do calculations with vectors just like ordinary numbers, as long as they are of the same length. Suppose that we also have the heights that correspond to the weights above. The body mass index (BMI) is defined for each person as the weight in kilograms divided by the square of the height in meters. This could be calculated as follows:

```
> height <- c(1.75, 1.80, 1.65, 1.90, 1.74, 1.91)
> bmi <- weight/height^2
> bmi
[1] 19.59184 22.22222 20.93664 24.93075 31.37799 19.73630
```

Notice that the operation is carried out elementwise (that is, the first value of `bmi` is $60/1.75^2$ and so forth) and that the `^` operator is used for raising a value to a power. (On some keyboards, `^` is a "dead key" and you will have to press the spacebar afterwards to make it show.)

It is in fact possible to perform arithmetic operations on vectors of different length. We already used that when we calculated the `height^2` part above since 2 has length 1. In such cases, the shorter vector is *recycled*. This is mostly used with vectors of length 1 (scalars) but sometimes also in other cases where a repeating pattern is desired. A warning is issued if the longer vector is not a multiple of the shorter in length.

These conventions for vectorized calculations make it very easy to specify typical statistical calculations. Consider, for instance, the calculation of the mean and standard deviation of the `weight` variable.

First, calculate the mean, $\bar{x} = \sum x_i / n$:

CALCULATING MEAN

```
> sum(weight)
[1] 446
> sum(weight)/length(weight)
[1] 74.33333
```

Then save the mean in a variable `xbar` and proceed with the calculation of $SD = \sqrt{(\sum(x_i - \bar{x})^2)/(n-1)}$. We do this in steps to see the individual components. The deviations from the mean are

```
> xbar <- sum(weight)/length(weight)
> weight - xbar
```

```
[1] -14.333333   -2.333333 -17.333333   15.666667  20.666667
[6]  -2.333333
```

Notice how `xbar`, which has length 1, is recycled and subtracted from each element of `weight`. The squared deviations will be

```
> (weight - xbar)^2
[1] 205.444444    5.444444 300.444444 245.444444 427.111111
[6]   5.444444
```

Since this command is quite similar to the one before it, it is convenient to enter it by editing the previous command. On most systems running R, the previous command can be recalled with the up-arrow key.

The sum of squared deviations is similarly obtained with

```
> sum((weight - xbar)^2)
[1] 1189.333
```

and all in all the standard deviation becomes

```
> sqrt(sum((weight - xbar)^2)/(length(weight) - 1))
[1] 15.42293
```

Of course, since R is a statistical program, such calculations are already built into the program, and you get the same results just by entering

```
> mean(weight)
[1] 74.33333
> sd(weight)
[1] 15.42293
```

1.1.4   Standard procedures

As a slightly more complicated example of what R can do, consider the following: The rule of thumb is that the BMI for a normal-weight individual should be between 20 and 25, and we want to know if our data deviate systematically from that. You might use a one-sample t test to assess whether the six persons' BMI can be assumed to have mean 22.5 given that they come from a normal distribution. To this end, you can use the function `t.test`. (You might not know the theory of the t test yet. The example is included here mainly to give some indication of what "real" statistical output looks like. A thorough description of `t.test` is given in Chapter 5.)

```
> t.test(bmi, mu=22.5)
          One Sample t-test
data:  bmi
t = 0.3449, df = 5, p-value = 0.7442
alternative hypothesis: true mean is not equal to 22.5
95 percent confidence interval:
 18.41734 27.84791
sample estimates:
mean of x
 23.13262
```

The argument mu=22.5 attaches a value to the formal argument mu, which represents the Greek letter μ conventionally used for the theoretical mean. If this is not given, t.test would use the default mu=0, which is not of interest here.

For a test like this, we get a more extensive printout than in the earlier examples. The details of the output are explained in Chapter 5, but you might focus on the *p*-value which is used for testing the hypothesis that the mean is 22.5. The *p*-value is not small, indicating that it is not at all unlikely to get data like those observed if the mean were in fact 22.5. (Loosely speaking; actually *p* is the probability of obtaining a *t* value bigger than 0.3449 or less than −0.3449.) However, you might also look at the 95% confidence interval for the true mean. This interval is quite wide, indicating that we really have very little information about the true mean.

1.1.5 Graphics

One of the most important aspects of the presentation and analysis of data is the generation of proper graphics. R — like S before it — has a model for constructing plots that allows simple production of standard plots as well as fine control over the graphical components.

If you want to investigate the relation between weight and height, the first idea is to plot one versus the other. This is done by

```
> plot(height,weight)
```

leading to Figure 1.2.

You will often want to modify the drawing in various ways. To that end, there are a wealth of plotting parameters that you can set. As an example, let us try changing the plotting symbol using the keyword pch ("plotting character") like this:

```
> plot(height, weight, pch=2)
```

same graph-diff markers

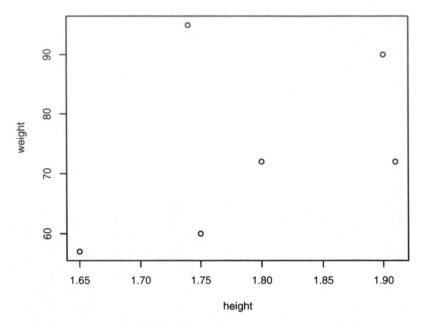

Figure 1.2. A simple *x–y* plot.

This gives the plot in Figure 1.3, with the points now marked with little triangles.

The idea behind the BMI calculation is that this value should be independent of the person's height, thus giving you a single number as an indication of whether someone is overweight and by how much. Since a normal BMI should be about 22.5, you would expect that *weight* ≈ $22.5 \times height^2$. Accordingly, you can superimpose a curve of expected weights at BMI 22.5 on the figure:

```
> hh <- c(1.65, 1.70, 1.75, 1.80, 1.85, 1.90)
> lines(hh, 22.5 * hh^2)
```

yielding Figure 1.4. The function lines will *add* (x,y) values joined by straight lines to an existing plot.

The reason for defining a new variable (hh) with heights rather than using the original height vector is twofold. First, the relation between height and weight is a quadratic one and hence nonlinear, although it can be difficult to see on the plot. Since we are approximating a nonlinear curve with a piecewise linear one, it will be better to use points that are spread evenly along the *x*-axis than to rely on the distribution of the original data. Second, since the values of `height` are not sorted, the line segments would not connect neighbouring points but would run back and forth between distant points.

1.2   R language essentials

This section outlines the basic aspects of the R language. It is necessary to do this in a slightly superficial manner, with some of the finer points glossed over. The emphasis is on items that are useful to know in interactive usage as opposed to actual programming, although a brief section on programming is included.

1.2.1   Expressions and objects

The basic interaction mode in R is one of expression evaluation. The user enters an expression; the system evaluates it and prints the result. Some expressions are evaluated not for their result but for *side effects* such as

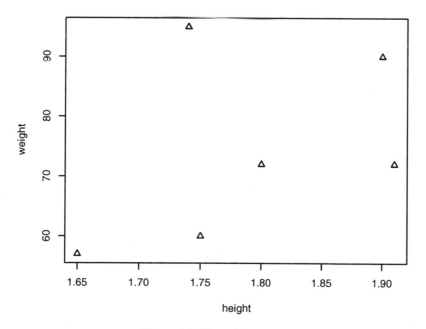

Figure 1.3. Plot with `pch` = 2.

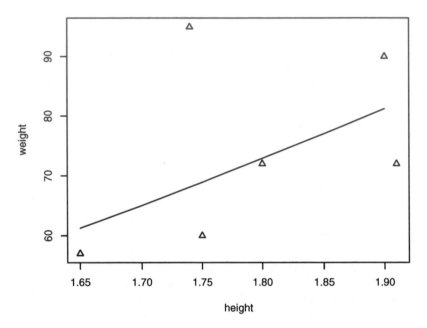

Figure 1.4. Superimposed reference curve, using `lines(...)`.

putting up a graphics window or writing to a file. All R expressions return a value (possibly `NULL`), but sometimes it is "invisible" and not printed.

Expressions typically involve variable references, operators such as +, and function calls, as well as some other items that have not been introduced yet.

Expressions work on *objects*. This is an abstract term for anything that can be assigned to a variable. R contains several different types of objects. So far, we have almost exclusively seen numeric vectors, but several other types are introduced in this chapter.

Although objects can be discussed abstractly, it would make a rather boring read without some indication of how to generate them and what to do with them. Conversely, much of the expression syntax makes little sense without knowledge of the objects on which it is intended to work. Therefore, the subsequent sections alternate between introducing new objects and introducing new language elements.

1.2.2   Functions and arguments

At this point, you have obtained an impression of the way R works, and
we have already used some of the special terminology when talking about
the plot *function*, etc. That is exactly the point: Many things in R are done
using *function calls*, commands that look like an application of a mathematical function of one or several variables; for example, `log(x)` or
`plot(height, weight)`.

The format is that a function name is followed by a set of parentheses containing one or more arguments. For instance, in `plot(height,weight)`
the function name is `plot` and the arguments are `height` and `weight`.
These are the *actual arguments*, which apply only to the current call. A function also has *formal arguments*, which get connected to actual arguments in
the call.

When you write `plot(height, weight)`, R assumes that the first argument corresponds to the *x*-variable and the second one to the *y*-variable.
This is known as *positional matching.* This becomes unwieldy if a function has a large number of arguments since you have to supply every
one of them and remember their position in the sequence. Fortunately,
R has methods to avoid this: Most arguments have sensible defaults and
can be omitted in the standard cases, and there are nonpositional ways of
specifying them when you need to depart from the default settings.

The `plot` function is in fact an example of a function that has a large
selection of arguments in order to be able to modify symbols, line
widths, titles, axis type, and so forth. We used the alternative form of
specifying arguments when setting the plot symbol to triangles with
`plot(height, weight, pch=2)`.

The `pch=2` form is known as a *named actual argument,* whose name can
be matched against the formal arguments of the function and thereby
allow *keyword matching* of arguments. The keyword `pch` was used to
say that the argument is a specification of the plotting character. This
type of function argument can be specified in arbitrary order. Thus, you
can write `plot(y=weight,x=height)` and get the same plot as with
`plot(x=height,y=weight)`.

The two kinds of argument specification — positional and named — can
be mixed in the same call.

Even if there are no arguments to a function call, you have to write, for
example, `ls()` for displaying the contents of the workspace. A common
error is to leave off the parentheses, which instead results in the display of
a piece of R code since `ls` entered by itself indicates that you want to see
the definition of the function rather than execute it.

The *formal arguments* of a function are part of the function definition. The set of formal arguments to a function, for instance `plot.default` (which is the function that gets called when you pass `plot` an x argument for which no special plot method exists), may be seen with

```
> args(plot.default)
function (x, y = NULL, type = "p", xlim = NULL, ylim = NULL,
    log = "", main = NULL, sub = NULL, xlab = NULL, ylab = NULL,
    ann = par("ann"), axes = TRUE, frame.plot = axes,
    panel.first = NULL, panel.last = NULL, asp = NA, ...)
```

Notice that most of the arguments have defaults, meaning that if you do not specify (say) the `type` argument, the function will behave as if you had passed `type="p"`. The `NULL` defaults for many of the arguments really serve as indicators that the argument is unspecified, allowing special behaviour to be defined inside the function. For instance, if they are not specified, the `xlab` and `ylab` arguments are constructed from the actual arguments passed as x and y. (There are some very fine points associated with this, but we do not go further into the topic.)

The triple-dot (. . .) argument indicates that this function will accept additional arguments of unspecified name and number. These are often meant to be passed on to other functions, although some functions treat it specially. For instance, in `data.frame` and `c`, the names of the . . .-arguments become the names of the elements of the result.

1.2.3   Vectors

We have already seen numeric vectors. There are two further types, character vectors and logical vectors.

A *character vector* is a vector of text strings, whose elements are specified and printed in quotes:

```
> c("Huey","Dewey","Louie")
[1] "Huey"  "Dewey" "Louie"
```

CHARACTER VECTORS!

It does not matter whether you use single- or double-quote symbols, as long as the left quote is the same as the right quote:

```
> c('Huey','Dewey','Louie')
[1] "Huey"  "Dewey" "Louie"
```

However, you should avoid the *acute accent* key (´), which is present on some keyboards. Double quotes are used throughout this book to prevent mistakes. Logical vectors can take the value `TRUE` or `FALSE` (or NA; see below). In input, you may use the convenient abbreviations `T` and `F` (if you are careful not to redefine them). Logical vectors are constructed using the c function just like the other vector types:

```
> c(T,T,F,T)
[1]   TRUE   TRUE FALSE   TRUE
```

LOGICAL VECTORS!

Actually, you will not often have to specify logical vectors in the manner above. It is much more common to use single logical values to turn an option on or off in a function call. Vectors of more than one value most often result from *relational expressions*:

```
> bmi > 25
[1] FALSE FALSE FALSE FALSE   TRUE FALSE
```

We return to relational expressions and logical operations in the context of conditional selection in Section 1.2.12.

1.2.4 *Quoting and escape sequences*   *lowKey confused*

Quoted character strings require some special considerations: How, for instance, do you put a quote symbol inside a string? And what about special characters such as newlines? This is done using *escape sequences*. We shall look at those in a moment, but first it will be useful to observe the following.

There is a distinction between a text string and the way it is printed. When, for instance, you give the string "Huey", it is a string of four characters, not six. The quotes are not actually part of the string, they are just there so that the system can tell the difference between a string and a variable name.

If you print a character vector, it usually comes out with quotes added to each element. There is a way to avoid this, namely to use the cat function. For instance,

```
> cat(c("Huey","Dewey","Louie"))
Huey Dewey Louie>
```

This prints the strings without quotes, just separated by a space character. There is no newline following the string, so the prompt (>) for the next line of input follows directly at the end of the line. (Notice that when the character vector is printed by cat there is no way of telling the difference from the single string "Huey Dewey Louie".)

To get the system prompt onto the next line, you must include a newline character

```
> cat("Huey","Dewey","Louie", "\n")
Huey Dewey Louie
>
```

Here, \n is an example of an escape sequence. It actually represents a single character, the linefeed (LF), but is represented as two. The backslash (\) is known as the *escape character*. In a similar vein, you can insert quote characters with \ ", as in

```
> cat("What is \"R\"?\n")
What is "R"?
```

There are also ways to insert other control characters and special glyphs, but it would lead us too far astray to discuss it in full detail. One important thing, though: What about the escape character itself? This, too, must be escaped, so to put a backslash in a string, you must double it. This is important to know when specifying file paths on Windows, see also Section 2.4.1.

1.2.5 Missing values

In practical data analysis, a data point is frequently unavailable (the patient did not show up, an experiment failed, etc.). Statistical software needs ways to deal with this. R allows vectors to contain a special NA value. This value is carried through in computations so that operations on NA yield NA as the result. There are some special issues associated with the handling of missing values; we deal with them as we encounter them (see "missing values" in the index).

1.2.6 Functions that create vectors

C, seq, rep

Here we introduce three functions, c, seq, and rep, that are used to create vectors in various situations.

The first of these, c, has already been introduced. It is short for "concatenate", joining items end to end, which is exactly what the function does:

```
> c(42,57,12,39,1,3,4)
[1] 42 57 12 39  1  3  4
```

You can also concatenate vectors of more than one element as in

```
> x <- c(1, 2, 3)
> y <- c(10, 20)
```

```
> c(x, y, 5)
[1]  1  2  3 10 20  5
```

However, you do not need to use c to create vectors of length 1. People sometimes type, for example, c(1), but it is the same as plain 1.

It is also possible to assign names to the elements. This modifies the way the vector is printed and is often used for display purposes.

```
> x <- c(red="Huey", blue="Dewey", green="Louie")
> x
    red     blue    green
 "Huey"  "Dewey"  "Louie"
```

(In this case, it *does* of course make sense to use c even for single-element vectors.)

The names can be extracted or set using names:

```
> names(x)
[1] "red"    "blue"   "green"
```

All elements of a vector have the same type. If you concatenate vectors of different types, they will be converted to the least "restrictive" type:

```
> c(FALSE, 3)
[1] 0 3
> c(pi, "abc")
[1] "3.14159265358979" "abc"
> c(FALSE, "abc")
[1] "FALSE" "abc"
```

That is, logical values may be converted to 0/1 or "FALSE"/"TRUE" and numbers converted to their printed representations.

The second function, seq ("sequence"), is used for equidistant series of numbers. Writing

```
> seq(4,9)
[1] 4 5 6 7 8 9
```

yields, as shown, the integers from 4 to 9. If you want a sequence in jumps of 2, write

```
> seq(4,10,2)
[1]  4  6  8 10
```

This kind of vector is frequently needed, particularly for graphics. For example, we previously used c(1.65,1.70,1.75,1.80,1.85,1.90) to define the x-coordinates for a curve, something that could also have been written seq(1.65,1.90,0.05) (the advantage of using seq might have been more obvious if the heights had been in steps of 1 cm rather than 5 cm!).

The case with step size equal to 1 can also be written using a special syntax:

```
> 4:9
[1] 4 5 6 7 8 9
```

The above is exactly the same as seq(4,9), only easier to read.

The third function, rep ("replicate"), is used to generate repeated values. It is used in two variants, depending on whether the second argument is a vector or a single number:

```
> oops <- c(7,9,13)
> rep(oops,3)
[1]   7  9 13   7  9 13   7  9 13
> rep(oops,1:3)
[1]   7  9  9 13 13 13
```

The first of the function calls above repeats the entire vector oops three times. The second call has the number 3 replaced by a vector with the three values (1, 2, 3); these values correspond to the elements of the oops vector, indicating that 7 should be repeated once, 9 twice, and 13 three times. The rep function is often used for things such as group codes: If it is known that the first 10 observations are men and the last 15 are women, you can use

```
> rep(1:2,c(10,15))
 [1] 1 1 1 1 1 1 1 1 1 1 2 2 2 2 2 2 2 2 2 2 2 2 2 2 2
```

to form a vector that for each observation indicates whether it is from a man or a woman.

The special case where there are equally many replications of each value can be obtained using the each argument. E.g., rep(1:2,each=10) is the same as rep(1:2,c(10,10)).

1.2.7   Matrices and arrays

A *matrix* in mathematics is just a two-dimensional array of numbers. Matrices are used for many purposes in theoretical and practical statistics, but it is not assumed that the reader is familiar with matrix algebra, so many special operations on matrices, including matrix multiplication, are skipped. (The document "An Introduction to R", which comes with the installation, outlines these items quite well.) However, matrices and also higher-dimensional arrays do get used for simpler purposes as well, mainly to hold tables, so an elementary description is in order.

In R, the matrix notion is extended to elements of any type, so you could have, for instance, a matrix of character strings. Matrices and arrays are represented as vectors with dimensions:

```
> x <- 1:12
> dim(x) <- c(3,4)                        3 × 4 dimension
> x
     [,1] [,2] [,3] [,4]
[1,]    1    4    7   10
[2,]    2    5    8   11
[3,]    3    6    9   12
```

The dim assignment function sets or changes the *dimension attribute* of x, causing R to treat the vector of 12 numbers as a 3 × 4 matrix. Notice that the storage is column-major; that is, the elements of the first column are followed by those of the second, etc.

A convenient way to create matrices is to use the matrix function:

```
> matrix(1:12,nrow=3,byrow=T)
     [,1] [,2] [,3] [,4]
[1,]    1    2    3    4        byrow =T
[2,]    5    6    7    8        flips rows from colums
[3,]    9   10   11   12
```

Notice how the byrow=T switch causes the matrix to be filled in a rowwise fashion rather than columnwise.

Useful functions that operate on matrices include rownames, colnames, and the transposition function t (notice the lowercase t as opposed to uppercase T for TRUE), which turns rows into columns and vice versa:

```
> x <- matrix(1:12,nrow=3,byrow=T)
> rownames(x) <- LETTERS[1:3]
> x
  [,1] [,2] [,3] [,4]
A    1    2    3    4
B    5    6    7    8
C    9   10   11   12
> t(x)
   A  B  C
[1,] 1  5  9
[2,] 2  6 10
[3,] 3  7 11
[4,] 4  8 12
```

flips

t (x)

transportation function

The character vector LETTERS is a built-in variable that contains the capital letters A–Z. Similar useful vectors are letters, month.name, and month.abb with lowercase letters, month names, and abbreviated month names.

You can "glue" vectors together, columnwise or rowwise, using the cbind and rbind functions.

```
> cbind(A=1:4,B=5:8,C=9:12)
     A B  C
[1,] 1 5  9
[2,] 2 6 10
[3,] 3 7 11
[4,] 4 8 12
> rbind(A=1:4,B=5:8,C=9:12)
  [,1] [,2] [,3] [,4]
A    1    2    3    4
B    5    6    7    8
C    9   10   11   12
```

cbind

rbind

We return to table operations in Section 4.5, which discusses tabulation of variables in a data set.

1.2.8   Factors

It is common in statistical data to have categorical variables, indicating some subdivision of data, such as social class, primary diagnosis, tumor stage, Tanner stage of puberty, etc. Typically, these are input using a numeric code.

Such variables should be specified as *factors* in R. This is a data structure that (among other things) makes it possible to assign meaningful names to the categories.

There are analyses where it is essential for R to be able to distinguish between categorical codes and variables whose values have a direct numerical meaning (see Chapter 7).

The terminology is that a factor has a set of *levels* — say four levels for concreteness. Internally, a four-level factor consists of two items: (a) a vector of integers between 1 and 4 and (b) a character vector of length 4 containing strings describing what the four levels are. Let us look at an example:

pain levels

factoring pain levels 0:3

```
> pain <- c(0,3,2,2,1)
> fpain <- factor(pain,levels=0:3)
> levels(fpain) <- c("none","mild","medium","severe")
```

The first command creates a numeric vector `pain`, encoding the pain levels of five patients. We wish to treat this as a categorical variable, so we create a factor `fpain` from it using the function `factor`. This is called with one argument in addition to `pain`, namely `levels=0:3`, which indicates that the *input* coding uses the values 0–3. The latter can in principle be left out since R by default uses the values in `pain`, suitably sorted, but it is a good habit to retain it; see below. The effect of the final line is that the level names are changed to the four specified character strings.

The result should be apparent from the following:

```
> fpain
[1] none    severe medium medium mild
Levels:   none mild medium severe
> as.numeric(fpain)
[1] 1 4 3 3 2
> levels(fpain)
[1] "none"   "mild"   "medium" "severe"
```

The function `as.numeric` extracts the numerical coding as numbers 1–4 and `levels` extracts the names of the levels. Notice that the original input coding in terms of numbers 0–3 has disappeared; the internal representation of a factor always uses numbers starting at 1.

R also allows you to create a special kind of factor in which the levels are ordered. This is done using the `ordered` function, which works similarly to `factor`. These are potentially useful in that they distinguish nominal and ordinal variables from each other (and arguably `text.pain` above ought to have been an ordered factor). Unfortunately, R defaults to treating the levels as if they were *equidistant* in the modelling code (by generating polynomial contrasts), so it may be better to ignore ordered factors at this stage.

1.2.9   Lists

It is sometimes useful to combine a collection of objects into a larger composite object. This can be done using *lists*.

You can construct a list from its components with the function `list`.

As an example, consider a set of data from Altman (1991, p. 183) concerning pre- and postmenstrual energy intake in a group of women. We can place these data in two vectors as follows:

```
> intake.pre <- c(5260,5470,5640,6180,6390,
+ 6515,6805,7515,7515,8230,8770)
> intake.post <- c(3910,4220,3885,5160,5645,
+ 4680,5265,5975,6790,6900,7335)
```

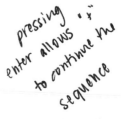
pressing "↵" enter allows you to continue the sequence

Notice how input lines can be broken and continue on the next line. If you press the Enter key while an expression is syntactically incomplete, R will assume that the expression continues on the next line and will change its normal > prompt to the *continuation prompt* +. This often happens inadvertently due to a forgotten parenthesis or a similar problem; in such cases, either complete the expression on the next line or press ESC (Windows and Macintosh) or Ctrl-C (Unix). The "Stop" button can also be used under Windows.

[handwritten margin note: counteract the "+"]

To combine these individual vectors into a list, you can say

```
> mylist <- list(before=intake.pre, after=intake.post)
> mylist
$before
 [1] 5260 5470 5640 6180 6390 6515 6805 7515 7515 8230 8770

$after
 [1] 3910 4220 3885 5160 5645 4680 5265 5975 6790 6900 7335
```

The components of the list are named according to the argument names used in `list`. Named components may be extracted like this:

```
> mylist$before
 [1] 5260 5470 5640 6180 6390 6515 6805 7515 7515 8230 8770
```

Many of R's built-in functions compute more than a single vector of values and return their results in the form of a list.

1.2.10   Data frames

A data frame corresponds to what other statistical packages call a "data matrix" or a "data set". It is a list of vectors and/or factors of the same length that are related "across" such that data in the same position come from the same experimental unit (subject, animal, etc.). In addition, it has a unique set of row names.

You can create data frames from preexisting variables:

```
> d <- data.frame(intake.pre, intake.post)
> d
  intake.pre intake.post
1       5260        3910
2       5470        4220
3       5640        3885
4       6180        5160
5       6390        5645
6       6515        4680
7       6805        5265
```

[handwritten margin note: same as fix(car)]

```
 8           7515        5975
 9           7515        6790
10           8230        6900
11           8770        7335
```

$ meaning

Notice that these data are paired, that is, the same woman has an intake of 5260 kJ premenstrually and 3910 kJ postmenstrually.

As with lists, components (i.e., individual variables) can be accessed using the $ notation:

```
> d$intake.pre
 [1] 5260 5470 5640 6180 6390 6515 6805 7515 7515 8230 8770
```

1.2.11 Indexing

If you need a particular element in a vector, for instance the premenstrual energy intake for woman no. 5, you can do

```
> intake.pre[5]
[1] 6390
```

The brackets are used for selection of data, also known as *indexing* or *sub-setting*. This also works on the left-hand side of an assignment (so that you can say, for instance, `intake.pre[5] <- 6390`) if you want to modify elements of a vector.

If you want a subvector consisting of data for more than one woman, for instance nos. 3, 5, and 7, you can index with a vector:

```
> intake.pre[c(3,5,7)]
[1] 5640 6390 6805
```

Note that it is necessary to use the `c(...)`-construction to define the vector consisting of the three numbers 3, 5, and 7. `intake.pre[3,5,7]` would mean something completely different. It would specify indexing into a three-dimensional array.

Of course, indexing with a vector also works if the index vector is stored in a variable. This is useful when you need to index several variables in the same way.

```
> v <- c(3,5,7)
> intake.pre[v]
[1] 5640 6390 6805
```

same way ... probs a bit longer

It is also worth noting that to get a sequence of elements, for instance the first five, you can use the a : b notation:

```
> intake.pre[1:5]
[1] 5260 5470 5640 6180 6390
```

A neat feature of R is the possibility of negative indexing. You can get all observations *except* nos. 3, 5, and 7 by writing

```
> intake.pre[-c(3,5,7)]
[1] 5260 5470 6180 6515 7515 7515 8230 8770
```

It is not possible to mix positive and negative indices. That would be highly ambiguous.

1.2.12 Conditional selection

We saw in Section 1.2.11 how to extract data using one or several indices. In practice, you often need to extract data that satisfy certain criteria, such as data from the males or the prepubertal or those with chronic diseases, etc. This can be done simply by inserting a relational expression instead of the index,

```
> intake.post[intake.pre > 7000]
[1] 5975 6790 6900 7335
```

yielding the postmenstrual energy intake for the four women who had an energy intake above 7000 kJ premenstrually.

Of course, this kind of expression makes sense only if the variables that go into the relational expression have the same length as the variable being indexed.

The comparison operators available are < (less than), > (greater than), == (equal to), <= (less than or equal to), >= (greater than or equal to), and != (not equal to). Notice that a double equal sign is used for testing equality. This is to avoid confusion with the = symbol used to match keywords with function arguments. Also, the != operator is new to some; the ! symbol indicates negation. The same operators are used in the C programming language.

To combine several expressions, you can use the logical operators & (logical "and"), | (logical "or"), and ! (logical "not"). For instance, we find the postmenstrual intake for women with a premenstrual intake between 7000 and 8000 kJ with

```
> intake.post[intake.pre > 7000 & intake.pre <= 8000]
[1] 5975 6790
```

There are also && and ||, which are used for flow control in R programming. However, their use is beyond what we discuss here.

It may be worth taking a closer look at what actually happens when you use a logical expression as an index. The result of the logical expression is a logical vector as described in Section 1.2.3:

```
> intake.pre > 7000 & intake.pre <= 8000
 [1] FALSE FALSE FALSE FALSE FALSE FALSE FALSE   TRUE   TRUE   FALSE
[11] FALSE
```

Indexing with a logical vector implies that you pick out the values where the logical vector is TRUE, so in the preceding example we got the 8th and 9th values in intake.post.

If missing values (NA; see Section 1.2.5) appear in an indexing vector, then R will create the corresponding elements in the result but set the values to NA.

In addition to the relational and logical operators, there are a series of functions that return a logical value. A particularly important one is is.na(x), which is used to find out which elements of x are recorded as missing (NA).

Notice that there is a real need for is.na because you cannot make comparisons of the form x==NA. That simply gives NA as the result for any value of x. The result of a comparison with an unknown value is unknown!

importance of "is.na"

1.2.13   Indexing of data frames

We have already seen how it is possible to extract variables from a data frame by typing, for example, d$intake.post. However, it is also possible to use a notation that uses the matrix-like structure directly:

```
> d <- data.frame(intake.pre,intake.post)
> d[5,1]
[1] 6390
```

gives fifth row, first column (that is, the "pre" measurement for woman no. 5), and

```
> d[5,]
  intake.pre intake.post
5       6390        5645
```

gives *all* measurements for woman no. 5. Notice that the comma in d[5,] is required; without the comma, for example d[2], you get the data frame consisting of the second *column* of d (that is, more like d[,2], which is the column itself).

Other indexing techniques also apply. In particular, it can be useful to extract all data for cases that satisfy some criterion, such as women with a premenstrual intake above 7000 kJ:

```
> d[d$intake.pre>7000,]
   intake.pre intake.post
8        7515        5975
9        7515        6790
10       8230        6900
11       8770        7335
```

Here we extracted the rows of the data frame where intake.pre>7000. Notice that the row names are those of the original data frame.

If you want to understand the details of this, it may be a little easier if it is divided into smaller steps. It could also have been done like this:

```
> sel <- d$intake.pre>7000
> sel
 [1] FALSE FALSE FALSE FALSE FALSE FALSE FALSE  TRUE  TRUE  TRUE
[11]  TRUE
> d[sel,]
   intake.pre intake.post
8        7515        5975
9        7515        6790
10       8230        6900
11       8770        7335
```

What happens is that sel *(select)* becomes a logical vector with the value TRUE for to the four women consuming more than 7000 kJ premenstrually. Indexing as d[sel,] yields data from the rows where sel is TRUE and from all columns because of the empty field after the comma.

It is often convenient to look at the first few cases in a data set. This can be done with indexing, like this:

```
> d[1:2,]
   intake.pre intake.post
1        5260        3910
2        5470        4220
```

This is such a frequent occurrence that a convenience function called head exists. By default, it shows the first six lines.

```
> head(d)
   intake.pre intake.post
1        5260        3910
2        5470        4220
```

3	5640	3885
4	6180	5160
5	6390	5645
6	6515	4680

Similarly, `tail` shows the last part.

1.2.14   Grouped data and data frames

The natural way of storing grouped data in a data frame is to have the data themselves in one vector and parallel to that have a factor telling which data are from which group. Consider, for instance, the following data set on energy expenditure for lean and obese women.

```
> energy
   expend stature
1    9.21   obese
2    7.53    lean
3    7.48    lean
4    8.08    lean
5    8.09    lean
6   10.15    lean
7    8.40    lean
8   10.88    lean
9    6.13    lean
10   7.90    lean
11  11.51   obese
12  12.79   obese
13   7.05    lean
14  11.85   obese
15   9.97   obese
16   7.48    lean
17   8.79   obese
18   9.69   obese
19   9.68   obese
20   7.58    lean
21   9.19   obese
22   8.11    lean
```

This is a convenient format since it generalizes easily to data classified by multiple criteria. However, sometimes it is desirable to have data in a separate vector for each group. Fortunately, it is easy to extract these from the data frame:

```
> exp.lean <- energy$expend[energy$stature=="lean"]
> exp.obese <- energy$expend[energy$stature=="obese"]
```

Alternatively, you can use the `split` function, which generates a list of vectors according to a grouping.

```
> l <- split(energy$expend, energy$stature)
> l
$lean
 [1]  7.53  7.48  8.08  8.09 10.15  8.40 10.88  6.13  7.90  7.05
[11]  7.48  7.58  8.11

$obese
[1]  9.21 11.51 12.79 11.85  9.97  8.79  9.69  9.68  9.19
```

1.2.15 Implicit loops

The looping constructs of R are described in Section 2.3.1. For the purposes
of this book, you can largely ignore their existence. However, there is a
group of R functions that it will be useful for you to know about.

A common application of loops is to apply a function to each element of
a set of values or vectors and collect the results in a single structure. In
R this is abstracted by the functions lapply and sapply. The former
always returns a list (hence the 'l'), whereas the latter tries to simplify
(hence the 's') the result to a vector or a matrix if possible. So, to compute
the mean of each variable in a data frame of numeric vectors, you can do
the following:

```
> lapply(thuesen, mean, na.rm=T)
$blood.glucose
[1] 10.3

$short.velocity
[1] 1.325652

> sapply(thuesen, mean, na.rm=T)
 blood.glucose short.velocity
      10.300000       1.325652
```

Notice how both forms attach meaningful names to the result, which
is another good reason to prefer to use these functions rather than explicit loops. The second argument to lapply/sapply is the function that
should be applied, here mean. Any further arguments are passed on to the
function; in this case we pass na.rm=T to request that missing values be
removed (see Section 4.1).

Sometimes you just want to repeat something a number of times but still
collect the results as a vector. Obviously, this makes sense only when the
repeated computations actually give different results, the common case
being simulation studies. This can be done using sapply, but there is a
simplified version called replicate, in which you just have to give a
count and the expression to evaluate:

```
> replicate(10,mean(rexp(20)))
 [1] 1.0677019 1.2166898 0.8923216 1.1281207 0.9636017 0.8406877
 [7] 1.3357814 0.8249408 0.9488707 0.5724575
```

A similar function, `apply`, allows you to apply a function to the rows or columns of a matrix (or over indices of a multidimensional array in general) as in

```
> m <- matrix(rnorm(12),4)
> m
            [,1]        [,2]         [,3]
[1,] -2.5710730 0.2524470 -0.16886795
[2,]  0.5509498 1.5430648  0.05359794
[3,]  2.4002722 0.1624704 -1.23407417
[4,]  1.4791103 0.9484525 -0.84670929
> apply(m, 2, min)
[1] -2.5710730  0.1624704 -1.2340742
```

The second argument is the index (or vector of indices) that defines what the function is applied to; in this case we get the columnwise minima.

Also, the function `tapply` allows you to create tables (hence the 't') of the value of a function on subgroups defined by its second argument, which can be a factor or a list of factors. In the latter case a cross-classified table is generated. (The grouping can also be defined by ordinary vectors. They will be converted to factors internally.)

```
> tapply(energy$expend, energy$stature, median)
 lean obese
 7.90  9.69
```

1.2.16 Sorting

It is trivial to sort a vector. Just use the `sort` function. (We use the built-in data set `intake` here; it contains the same data that were used in Section 1.2.9.)

```
> intake$post
 [1] 3910 4220 3885 5160 5645 4680 5265 5975 6790 6900 7335
> sort(intake$post)
 [1] 3885 3910 4220 4680 5160 5265 5645 5975 6790 6900 7335
```

(`intake$pre` could not be used for this example since it is sorted already!)

However, sorting a single vector is not always what is required. Often you need to sort a series of variables according to the values of some *other* variables — blood pressures sorted by sex and age, for instance. For this purpose, there is a construction that may look somewhat abstract at first but is really very powerful. You first compute an *ordering* of a variable.

```
> order(intake$post)
 [1]  3  1  2  6  4  7  5  8  9 10 11
```

The result is the numbers 1 to 11 (or whatever the length of the vector is), sorted according to the size of the argument to `order` (here `intake$post`). Interpreting the result of `order` is a bit tricky — it should be read as follows: You sort `intake$post` by placing its values in the order no. 3, no. 1, no. 2, no. 6, etc.

The point is that, by indexing with this vector, other variables can be sorted by the same criterion. Note that indexing with a vector containing the numbers from 1 to the number of elements exactly once corresponds to a reordering of the elements.

```
> o <- order(intake$post)
> intake$post[o]
 [1] 3885 3910 4220 4680 5160 5265 5645 5975 6790 6900 7335
> intake$pre[o]
 [1] 5640 5260 5470 6515 6180 6805 6390 7515 7515 8230 8770
```

What has happened here is that `intake$post` has been sorted — just as in `sort(intake$post)` — while `intake$pre` has been sorted by the size of the corresponding `intake$post`.

It is of course also possible to sort the entire data frame `intake`

```
> intake.sorted <- intake[o,]
```

Sorting by several criteria is done simply by having several arguments to `order`; for instance, `order(sex,age)` will give a main division into men and women, and within each sex an ordering by age. The second variable is used when the order cannot be decided from the first variable. Sorting in reverse order can be handled by, for example, changing the sign of the variable.

1.3 Exercises

1.1 How would you check whether two vectors are the same if they may contain missing (NA) values? (Use of the `identical` function is considered cheating!)

1.2 If x is a factor with n levels and y is a length n vector, what happens if you compute `y[x]`?

1.3 Write the logical expression to use to extract girls between 7 and 14 years of age in the `juul` data set.

1.4 What happens if you change the levels of a factor (with `levels`) and give the same value to two or more levels?

1.5 On p. 27, `replicate` was used to simulate the distribution of the mean of 20 random numbers from the exponential distribution by repeating the operation 10 times. How would you do the same thing with `sapply`?

2

The R environment

This chapter collects some practical aspects of working with R. It describes issues regarding the structure of the workspace, graphical devices and their parameters, and elementary programming, and includes a fairly extensive, although far from complete, discussion of data entry.

2.1 Session management

2.1.1 The workspace

All variables created in R are stored in a common workspace. To see which variables are defined in the workspace, you can use the function ls (*list*). It should look as follows if you have run all the examples in the preceding chapter:

```
> ls()
 [1] "bmi"            "d"              "exp.lean"
 [4] "exp.obese"      "fpain"          "height"
 [7] "hh"             "intake.post"    "intake.pre"
[10] "intake.sorted"  "l"              "m"
[13] "mylist"         "o"              "oops"
[16] "pain"           "sel"            "v"
[19] "weight"         "x"              "xbar"
[22] "y"
```

P. Dalgaard, *Introductory Statistics with R*,
DOI: 10.1007/978-0-387-79054-1_2, © Springer Science+Business Media, LLC 2008

Remember that you cannot omit the parentheses in `ls()`.

If at some point things begin to look messy, you can delete some of the objects. This is done using `rm` (*remove*), so that   *deletes the variables*

```
> rm(height, weight)
```

deletes the variables `height` and `weight`.

The entire workspace can be cleared using `rm(list=ls())` and also via the "Remove all objects" or "Clear Workspace" menu entries in the Windows and Macintosh GUIs. This does not remove variables whose name begins with a dot because they are not listed by `ls()` — you would need `ls(all=T)` for that, but it could be dangerous because such names are used for system purposes.

If you are acquainted with the Unix operating system, for which the S language, which preceded R, was originally written, then you will know that the commands for listing and removing files in Unix are called precisely `ls` and `rm`.

It is possible to save the workspace to a file at any time. If you just write

```
save.image()
```

then it will be saved to a file called `.RData` in your working directory. The Windows version also has this on the File menu. When you exit R, you will be asked whether to save the workspace image; if you accept, the same thing will happen. It is also possible to specify an alternative filename (within quotes). You can also save selected objects with `save`. The `.RData` file is loaded by default when R is started in its directory. Other save files can be loaded into your current workspace using `load`.

Save the workspace!

2.1.2   Textual output

It is important to note that the workspace consists only of R objects, not of any of the output that you have generated during a session. If you want to save your output, use "Save to File" from the File menu in Windows or use standard cut-and-paste facilities. You can also use ESS (Emacs Speaks Statistics), which works on all platforms. It is a "mode" for the Emacs editor where you can run your entire session in an Emacs buffer. You can get ESS and installation instructions for it from CRAN (see Appendix A).

An alternative way of diverting output to a file is to use the `sink` function. This is largely a relic from the days of the 80×25 computer terminal, where cut-and-paste techniques were not available, but it can still be useful at times. In particular, it can be used in batch processing. The way it works is as follows:

```
> sink("myfile")
> ls()
```

no output!

No output appears! <u>This is because the output goes into the file</u> `myfile` in <u>the current directory</u>. The system will remain in a state where commands are processed, but the output (apparently) goes into the drain until the normal state of affairs is reestablished by

```
> sink()
```

The current working directory can be obtained by `getwd()` and changed by `setwd(mydir)`, where `mydir` is a character string. The initial working directory is system-dependent; for instance, the Windows GUI sets it to the user's home directory, and command line versions use the directory from which you start R.

2.1.3 Scripting

Beyond a certain level of complexity, you will not want to work with R on a line-by-line basis. For instance, if you have entered an 8 × 8 matrix over eight lines and realize that you made a mistake, you will find yourself using the up-arrow key 64 times to reenter it! In such cases, it is better to work with R *scripts*, collections of lines of R code stored either in a file or in computer memory somehow.

One option is to use the `source` function, which is sort of the opposite of `sink`. It takes the input (i.e., the commands from a file) and runs them. Notice, though, that the entire file is syntax-checked before anything is executed. It is often useful to set `echo=T` in the call so that commands are printed along with the output.

Another option is more interactive in nature. You can work with a script editor window, which allows you to submit one or more lines of the script to a running R, which will then behave as if the same lines had been entered at the prompt. The Windows and Macintosh versions of R have simple scripting windows built-in, and a number of text editors also have features for sending commands to R; popular choices on Windows include TINN-R and WinEdt. This is also available as part of ESS (see the preceding section).

The history of commands entered in a session can be saved and reloaded using the <u>savehistory</u> and <u>loadhistory</u> commands, which are also mapped to menu entries in Windows. Saved histories can be useful as a starting point for writing scripts; notice also that the `history()` function will show the last commands entered at the console (up to a maximum of 25 lines by default).

2.1.4   Getting help

R can do a lot more than what a typical beginner can be expected to need or even understand. This book is written so that most of the code you are likely to need in relation to the statistical procedures is described in the text, and the compendium in Appendix C is designed to provide a basic overview. However, it is obviously not possible to cover everything.

R also comes with extensive online help in text form as well as in the form of a series of HTML files that can be read using a Web browser such as Netscape or Internet Explorer. The help pages can be accessed via "help" in the menu bar on Windows and by entering `help.start()` on any platform. You will find that the pages are of a technical nature. Precision and conciseness here take precedence over readability and pedagogy (something one learns to appreciate after exposure to the opposite).

From the command line, you can always enter `help(aggregate)` to get help on the `aggregate` function or use the prefix form `?aggregate`. If the HTML viewer is running, then the help page is shown there. Otherwise it is shown as text either through a pager to the terminal window or in a separate window.

Notice that the HTML version of the help system features a very useful "Search Engine and Keywords" and that the `apropos` function allows you to get a list of command names that contain a given pattern. The function `help.search` is similar but uses fuzzy matching and searches deeper into the help pages, so that it will be able to locate, for example, Kendall's correlation coefficient in `cor.test` if you use `help.search("kendal")`.

Also available with the R distributions is a set of documents in various formats. Of particular interest is "An Introduction to R", originally based on a set of notes for S-PLUS by Bill Venables and David Smith and modified for R by various people. It contains an introduction to the R language and environment in a rather more language-centric fashion than this book. On the Windows platform, you can choose to install PDF documents as part of the installation procedure so that — provided the Adobe Acrobat Reader program is also installed — it can be accessed via the Help menu. An HTML version (without pictures) can be accessed via the browser interface on all platforms.

2.1.5   Packages

An R installation contains one or more libraries of packages. Some of these packages are part of the basic installation. Others can be downloaded from CRAN (see Appendix A), which currently hosts over 1000 packages for various purposes. You can even create your own packages.

A library is generally just a folder on your disk. A system library is created when R is installed. In some installations, users may be prohibited from modifying the system library. It is possible to set up private user libraries; see help(".Library") for details.

A package can contain functions written in the R language, dynamically loaded libraries of compiled code (written in C or Fortran mostly), and data sets. It generally implements functionality that most users will probably not need to have loaded all the time. A package is loaded into R using the library command, so to load the survival package you should enter

```
> library(survival)
```

The loaded packages are not considered part of the user workspace. If you terminate your R session and start a new session with the saved workspace, then you will have to load the packages again. For the same reason, it is rarely necessary to remove a package that you have loaded, but it can be done if desired with

```
> detach("package:survival")
```

(see also Section 2.1.7).

2.1.6   Built-in data

Many packages, both inside and outside the standard R distribution, come with built-in data sets. Such data sets can be rather large, so it is not a good idea to keep them all in computer memory at all times. A mechanism for on-demand loading is required. In many packages, this works via a mechanism called *lazy loading*, which allows the system to "pretend" that the data are in memory, but in fact they are not loaded until they are referenced for the first time.

With this mechanism, data are "just there". For example, if you type "thuesen", the data frame of that name is displayed. Some packages still require explicit calls to the data function. Most often, this loads a data frame with the name that its argument specifies; data(thuesen) will, for instance, load the thuesen data frame.

What `data` does is to go through the data directories associated with each package (see Section 2.1.5) and look for files whose basename matches the given name. Depending on the file extension, several things can then happen. Files with a `.tab` extension are read using `read.table` (Section 2.4), whereas files with a `.R` extension are executed as source files (and could, in general, do anything!), to give two common examples.

If there is a subdirectory of the current directory called `data`, then it is searched as well. This can be quite a handy way of organizing your personal projects.

2.1.7 *attach and detach*

The notation for accessing variables in data frames gets rather heavy if you repeatedly have to write longish commands like

```
plot(thuesen$blood.glucose,thuesen$short.velocity)
```

Fortunately, you can make R look for objects among the variables in a given data frame, for example `thuesen`. You write

```
> attach(thuesen)
```

and then `thuesen`'s data are available without the clumsy $-notation:

```
> blood.glucose
 [1] 15.3 10.8  8.1 19.5  7.2  5.3  9.3 11.1  7.5 12.2  6.7  5.2
[13] 19.0 15.1  6.7  8.6  4.2 10.3 12.5 16.1 13.3  4.9  8.8  9.5
```

What happens is that the data frame `thuesen` is placed in the system's *search path*. You can view the search path with `search`:

```
> search()
 [1] ".GlobalEnv"        "thuesen"            "package:ISwR"
 [4] "package:stats"     "package:graphics"   "package:grDevices"
 [7] "package:utils"     "package:datasets"   "package:methods"
[10] "Autoloads"         "package:base"
```

Notice that `thuesen` is placed as no. 2 in the search path. `.GlobalEnv` is the workspace and `package:base` is the system library where all standard functions are defined. `Autoloads` is not described here. `package:stats` and onwards contains the basic statistical routines such as the Wilcoxon test, and the other packages similarly contain various functions and data sets. (The package system is modular, and you can run R with a minimal set of packages for specific uses.) Finally, `package:ISwR` contains the data sets used for this book.

There may be several objects of the same name in different parts of the search path. In that case, R chooses the first one (that is, it searches first in .GlobalEnv, then in thuesen, and so forth). For this reason, you need to be a little careful with "loose" objects that are defined in the workspace outside a data frame since they will be used before any vectors and factors of the same name in an attached data frame. For the same reason, it is not a good idea to give a data frame the same name as one of the variables inside it. Note also that changing a data frame after attaching it will not affect the variables available since attach involves a (virtual) copy operation of the data frame.

It is not possible to attach data frames in front of .GlobalEnv or following package:base. However, it is possible to attach more than one data frame. New data frames are inserted into position 2 by default, and everything except .GlobalEnv moves one step to the right. It is, however, possible to specify that a data frame should be searched before .GlobalEnv by using constructions of the form

```
with(thuesen, plot(blood.glucose, short.velocity))
```

In some contexts, R uses a slightly different method when looking for objects. If looking for a variable of a specific type (usually a function), R will skip those of other types. This is what saves you from the worst consequences of accidentally naming a variable (say) c, even though there is a system function of the same name.

You can remove a data frame from the search path with detach. If no arguments are given, the data frame in position 2 is removed, which is generally what is desired. .GlobalEnv and package:base cannot be detach'ed.

```
> detach()
> search()
 [1] ".GlobalEnv"        "package:ISwR"      "package:stats"
 [4] "package:graphics"  "package:grDevices" "package:utils"
 [7] "package:datasets"  "package:methods"   "Autoloads"
[10] "package:base"
```

2.1.8 *subset,* transform, *and* within

You can attach a data frame to avoid the cumbersome indexing of every variable inside of it. However, this is less helpful for selecting subsets of data and for creating new data frames with transformed variables. A couple of functions exist to make these operations easier. They are used as follows:

```
> thue2 <- subset(thuesen,blood.glucose<7)
> thue2
   blood.glucose short.velocity
6            5.3           1.49
11           6.7           1.25
12           5.2           1.19
15           6.7           1.52
17           4.2           1.12
22           4.9           1.03
> thue3 <- transform(thuesen,log.gluc=log(blood.glucose))
> thue3
   blood.glucose short.velocity log.gluc
1           15.3           1.76 2.727853
2           10.8           1.34 2.379546
3            8.1           1.27 2.091864
4           19.5           1.47 2.970414
5            7.2           1.27 1.974081
...
22           4.9           1.03 1.589235
23           8.8           1.12 2.174752
24           9.5           1.70 2.251292
```

Notice that the variables used in the expressions for new variables or for subsetting are evaluated with variables taken from the data frame.

`subset` also works on single vectors. This is nearly the same as indexing with a logical vector (such as `short.velocity[blood.glucose<7]`), except that observations with missing values in the selection criterion are excluded.

`subset` also has a `select` argument which can be used to extract variables from the data frame. We shall return to this in Section 10.3.1.

The `transform` function has a couple of drawbacks, the most serious of which is probably that it does not allow chained calculations where some of the new variables depend on the others. The = signs in the syntax are not assignments, but indicate names, which are assigned to the computed vectors in the last step.

An alternative to `transform` is the `within` function, which can be used like this:

```
> thue4 <- within(thuesen, {
+     log.gluc <- log(blood.glucose)
+     m <- mean(log.gluc)
+     centered.log.gluc <- log.gluc - m
+     rm(m)
+ })
> thue4
   blood.glucose short.velocity centered.log.gluc log.gluc
1           15.3           1.76       0.481879807 2.727853
2           10.8           1.34       0.133573113 2.379546
```

3	8.1	1.27	−0.154108960	2.091864
4	19.5	1.47	0.724441444	2.970414
5	7.2	1.27	−0.271891996	1.974081
...				
22	4.9	1.03	−0.656737817	1.589235
23	8.8	1.12	−0.071221300	2.174752
24	9.5	1.70	0.005318777	2.251292

Notice that the second argument is an arbitrary expression (here a *compound* expression, see p. 45). The function is similar to `with`, but instead of just returning the computed value, it collects all new and modified variables into a modified data frame, which is then returned. As shown, variables containing intermediate results can be discarded with `rm`. (It is particularly important to do this if the contents are incompatible with the data frame.)

2.2 The graphics subsystem

In Section 1.1.5, we saw how to generate a simple plot and superimpose a curve on it. It is quite common in statistical graphics for you to want to create a plot that is slightly different from the default: Sometimes you will want to add annotation, sometimes you want the axes to be different — labels instead of numbers, irregular placement of tick marks, etc. All these things can be obtained in R. The methods for doing them may feel slightly unusual at first, but offers a very flexible and powerful approach.

In this section, we look deeper into the structure of a typical plot and give some indication of how you can work with plots to achieve your desired results. Beware, though, that this is a large and complex area and it is not within the scope of this book to cover it completely. In fact, we completely ignore important newer tools in the `grid` and `lattice` packages.

2.2.1 Plot layout

In the graphics model that R uses, there is (for a single plot) a figure region containing a central plotting region surrounded by margins. Coordinates inside the plotting region are specified in data units (the kind generally used to label the axes). Coordinates in the margins are specified in *lines of text* as you move in a direction perpendicular to a side of the plotting region but in data units as you move along the side. This is useful since you generally want to put text in the margins of a plot.

A standard *x–y* plot has an *x* and a *y* title label generated from the expressions being plotted. You may, however, override these labels and also add two further titles, a main title above the plot and a subtitle at the very bottom, in the plot call.

```
> x <- runif(50,0,2)
> y <- runif(50,0,2)
> plot(x, y, main="Main title", sub="subtitle",
+        xlab="x-label", ylab="y-label")
```

Inside the plotting region, you can place points and lines that are either specified in the `plot` call or added later with `points` and `lines`. You can also place a text with

```
> text(0.6,0.6,"text at (0.6,0.6)")
> abline(h=.6,v=.6)
```

Here, the `abline` call is just to show how the text is centered on the point $(0.6, 0.6)$. (Normally, `abline` plots the line $y = a + bx$ when given a and b as arguments, but it can also be used to draw horizontal and vertical lines as shown.)

The margin coordinates are used by the `mtext` function. They can be demonstrated as follows:

```
> for (side in 1:4) mtext(-1:4,side=side,at=.7,line=-1:4)
> mtext(paste("side",1:4), side=1:4, line=-1,font=2)
```

The `for` loop (see Section 2.3.1) places the numbers -1 to 4 on corresponding lines in each of the four margins at an off-center position of 0.7 measured in user coordinates. The subsequent call places a label on each side, giving the side number. The argument `font=2` means that a boldface font is used. Notice in Figure 2.1 that not all the margins are wide enough to hold all the numbers and that it is possible to use negative line numbers to place text within the plotting region.

2.2.2 *Building a plot from pieces*

High-level plots are composed of elements, each of which can also be drawn separately. The separate drawing commands often allow finer control of the element, so a standard strategy to achieve a given effect is first to draw the plot without that element and add the element subsequently. As an extreme case, the following command will plot absolutely nothing:

```
> plot(x, y, type="n", xlab="", ylab="", axes=F)
```

Here `type="n"` causes the points not to be drawn. `axes=F` suppresses the axes and the box around the plot, and the x and y title labels are set to empty strings.

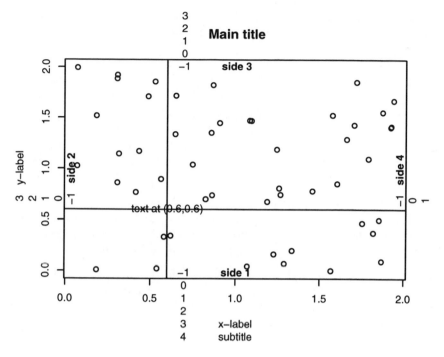

Figure 2.1. The layout of a standard plot.

However, the fact that nothing is plotted does not mean that nothing happened. The command sets up the plotting region and coordinate systems just as if it had actually plotted the data. To add the plot elements, evaluate the following:

```
> points(x,y)
> axis(1)
> axis(2,at=seq(0.2,1.8,0.2))
> box()
> title(main="Main title", sub="subtitle",
+     xlab="x-label", ylab="y-label")
```

Notice how the second `axis` call specifies an alternative set of tick marks (and labels). This is a common technique used to create special axes on a plot and might also be used to create nonequidistant axes as well as axes with nonnumeric labelling.

Plotting with `type="n"` is sometimes a useful technique because it has the side effect of dimensioning the plot area. For instance, to create a plot with different colours for different groups, you could first plot all data with `type="n"`, ensuring that the plot region is large enough, and then add the points for each group using `points`. (Passing a vector argument for `col` is more expedient in this particular case.)

2.2.3 Using `par`

The `par` function allows incredibly fine control over the details of a plot, although it can be quite confusing to the beginner (and even to experienced users at times). The best strategy for learning it may well be simply to try and pick up a few useful tricks at a time and once in a while try to solve a particular problem by poring over the help page.

Some of the parameters, but not all, can also be set via arguments to plotting functions, which also have some arguments that cannot be set by `par`. When a parameter can be set by both methods, the difference is generally that if something is set via `par`, then it stays set subsequently.

The `par` settings allow you to control line width and type, character size and font, colour, style of axis calculation, size of the plot and figure regions, clipping, etc. It is possible to divide a figure into several subfigures by using the `mfrow` and `mfcol` parameters.

For instance, the default margin sizes are just over 5, 4, 4, and 2 lines. You might set `par(mar=c(4,4,2,2)+0.1)` before plotting. This shaves one line off the bottom margin and two lines off the top margin of the plot, which will reduce the amount of unused whitespace when there is no main title or subtitle. If you look carefully, you will in fact notice that Figure 2.1 has a somewhat smaller plotting region than the other plots in this book. This is because the other plots have been made with reduced margins for typesetting reasons.

However, it is quite pointless to describe the graphics parameters completely at this point. Instead, we return to them as they are used for specific plots.

2.2.4 Combining plots

Some special considerations arise when you wish to put several elements together in the same plot. Consider overlaying a histogram with a normal density (see Sections 4.2 and 4.4.1 for information on histograms and Section 3.5.1 for density). The following is close, but only nearly good enough (figure not shown).

```
> x <- rnorm(100)
> hist(x,freq=F)
> curve(dnorm(x),add=T)
```

The `freq=F` argument to `hist` ensures that the histogram is in terms of densities rather than absolute counts. The `curve` function graphs an expression (in terms of x) and its `add=T` allows it to overplot an existing plot. So things are generally set up correctly, but sometimes the top of the density function gets chopped off. The reason is of course that the height of the normal density played no role in the setting of the y-axis for the histogram. It will not help to reverse the order and draw the curve first and add the histogram because then the highest bars might get clipped.

The solution is first to get hold of the magnitude of the y values for both plot elements and make the plot big enough to hold both (Figure 2.2):

```
> h <- hist(x, plot=F)
> ylim <- range(0, h$density, dnorm(0))
> hist(x, freq=F, ylim=ylim)
> curve(dnorm(x), add=T)
```

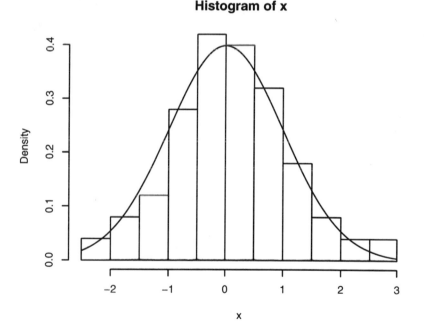

Figure 2.2. Histogram with normal density overlaid.

When called with `plot=F`, `hist` will not plot anything, but it will return a structure containing the bar heights on the density scale. This and the fact that the maximum of `dnorm(x)` is `dnorm(0)` allows us to calculate a range covering both the bars and the normal density. The zero in the `range` call ensures that the bottom of the bars will be in range, too. The range of *y* values is then passed to the `hist` function via the `ylim` argument.

2.3   R programming

It is possible to write your own R functions. In fact, this is a major aspect and attraction of working with the system in the long run. This book largely avoids the issue in favour of covering a larger set of basic statistical procedures that can be executed from the command line. However, to give you a feel for what can be done, consider the following function, which wraps the code from the example of Section 2.2.4 so that you can just say `hist.with.normal(rnorm(200))`. It has been slightly extended so that it now uses the empirical mean and standard deviation of the data instead of just 0 and 1.

```
> hist.with.normal <- function(x, xlab=deparse(substitute(x)),...)
+ {
+      h <- hist(x, plot=F, ...)
+      s <- sd(x)
+      m <- mean(x)
+      ylim <- range(0,h$density,dnorm(0,sd=s))
+      hist(x, freq=F, ylim=ylim, xlab=xlab, ...)
+      curve(dnorm(x,m,s), add=T)
+ }
```

Notice the use of a default argument for `xlab`. If `xlab` is not specified, then it is obtained from this expression, which evaluates to a character form of the expression given for *x*; that is, if you pass `rnorm(100)` for *x*, then the *x* label becomes "rnorm(100)". Notice also the use of a `...` argument, which collects any additional arguments and passes them on to `hist` in the two calls.

You can learn more about programming in R by studying the built-in functions, starting with simple ones like `log10` or `weighted.mean`.

2.3.1   Flow control

Until now, we have seen components of the R language that cause evaluation of single expressions. However, R is a true programming language that allows conditional execution and looping constructs as well. Consider, for instance, the following code. (The code implements a version of Newton's method for calculating the square root of y.)

```
> y <- 12345
> x <- y/2
> while (abs(x*x-y) > 1e-10) x <- (x + y/x)/2
> x
[1] 111.1081
> x^2
[1] 12345
```

Notice the `while(condition) expression` construction, which says that the expression should be evaluated as long as the condition is TRUE. The test occurs at the top of the loop, so the expression might never be evaluated.

A variation of the same algorithm with the test at the bottom of the loop can be written with a `repeat` construction:

```
> x <- y/2
> repeat{
+       x <- (x + y/x)/2
+       if (abs(x*x-y) < 1e-10) break
+ }
> x
[1] 111.1081
```

This also illustrates three other flow control structures: (a) a *compound expression*, several expressions held together between curly braces; (b) an `if` construction for conditional execution; and (c) a `break` expression, which causes the enclosing loop to exit.

Incidentally, the loop could allow for y being a vector simply by changing the termination condition to

```
    if (all(abs(x*x - y) < 1e-10)) break
```

This would iterate excessively for some elements, but the vectorized arithmetic would likely more than make up for that.

However, the most frequently used looping construct is `for`, which loops over a fixed set of values as in the following example, which plots a set of power curves on the unit interval.

```
> x <- seq(0, 1,.05)
> plot(x, x, ylab="y", type="l")
> for ( j in 2:8 ) lines(x, x^j)
```

Notice the *loop variable* j, which in turn takes the values of the given sequence when used in the `lines` call.

2.3.2   Classes and generic functions

Object-oriented programming is about creating coherent systems of data
and methods that work upon them. One purpose is to simplify programs
by accommodating the fact that you will have conceptually similar methods for different types of data, even though the implementations will have
to be different. A prototype example is the print method: It makes sense
to print many kinds of data objects, but the print layout will depend on
what the data object is. You will generally have a *class* of data objects and
a *print method* for that class. There are several object-oriented languages
implementing these ideas in different ways.

Most of the basic parts of R use the same object system as S version 3. An
alternative object system similar to that of S version 4 has been developed
in recent years. The new system has several advantages over the old one,
but we shall restrict attention to the latter. The S3 object system is a simple system in which an object has a `class` attribute, which is simply a
character vector. One example of this is that all the return values of the
classical tests such as `t.test` have class `"htest"`, indicating that they
are the result of a hypothesis test. When these objects are printed, it is
done by `print.htest`, which creates the nice layout (see Chapter 5 for
examples). However, from a programmatic viewpoint, these objects are
just lists, and you can, for instance, extract the *p*-value by writing

```
> t.test(bmi, mu=22.5)$p.value
[1] 0.7442183
```

The function `print` is a *generic function*, one that acts differently depending on its argument. These generally look like this:

```
> print
function (x, ...)
UseMethod("print")
<environment: namespace:base>
```

What `UseMethod("print")` means is that R should pass control to a
function named according to the object class (`print.htest` for objects of
class `"htest"`, etc.) or, if this is not found, to `print.default`. To see all
the methods available for `print`, type `methods(print)` (there are 138
of them in R 2.6.2, so the output is not shown here).

2.4   Data entry

Data sets do not have to be very large before it becomes impractical to type
them in with `c(...)`. Most of the examples in this book use data sets in-

cluded in the ISwR package, made available to you by library(ISwR). However, as soon as you wish to apply the methods to your own data, you will have to deal with data file formats and the specification thereof.

In this section we discuss how to read data files and how to use the data editor module in R. The text has some bias toward Windows systems, mainly because of some special issues that need to be mentioned for that platform.

2.4.1 Reading from a text file

The most convenient way of reading data into R is via the function called read.table. It requires that data be in "ASCII format"; that is, a "flat file" as created with Windows' NotePad or any plain-text editor. The result of read.table is a data frame, and it expects to find data in a corresponding layout where each line in the file contains all data from one subject (or rat or ...) in a specific order, separated by blanks or, optionally, some other separator. The first line of the file can contain a header giving the names of the variables, a practice that is highly recommended.

Table 11.6 in Altman (1991) contains an example on ventricular circumferential shortening velocity versus fasting blood glucose by Thuesen et al. We used those data to illustrate subsetting and use them again in the chapter on correlation and regression. They are among the built-in data sets in the ISwR package and available as the data frame thuesen, but the point here is to show how to read them from a plain-text file.

Assume that the data are contained in the file thuesen.txt, which looks as follows:

```
blood.glucose    short.velocity
15.3             1.76
10.8             1.34
8.1              1.27
19.5             1.47
7.2              1.27
5.3              1.49
9.3              1.31
11.1             1.09
7.5              1.18
12.2             1.22
6.7              1.25
5.2              1.19
19.0             1.95
15.1             1.28
6.7              1.52
8.6              NA
4.2              1.12
```

```
10.3            1.37
12.5            1.19
16.1            1.05
13.3            1.32
4.9             1.03
8.8             1.12
9.5             1.70
```

To enter the data into the file, you could start up Windows' NotePad or any other plain-text editor, such as those discussed in Section 2.1.3. Unix/Linux users should just use a standard editor, such as emacs or vi. If you must, you can even use a word processing program with a little care.

You should simply type in the data as shown. Notice that the columns are separated by an arbitrary number of blanks and that NA represents a missing value.

At the end, you should save the data to a text file. Notice that word processors require special actions in order to save as text. Their normal save format is difficult to read from other programs.

Assuming further that the file is in the ISwR folder on the N: drive, the data can be read using

```
> thuesen2 <- read.table("N:/ISwR/thuesen.txt",header=T)
```

Notice header=T specifying that the first line is a header containing the names of variables contained in the file. Also note that you use forward slashes (/), not backslashes (\), in the filename, even on a Windows system.

The reason for avoiding backslashes in Windows filenames is that the symbol is used as an escape character (see Section 1.2.4) and therefore needs to be doubled. You could have used N:\\ISwR\\thuesen.txt.

The result is a data frame, which is assigned to the variable thuesen2 and looks as follows:

```
> thuesen2
   blood.glucose short.velocity
1          15.3           1.76
2          10.8           1.34
3           8.1           1.27
4          19.5           1.47
5           7.2           1.27
6           5.3           1.49
7           9.3           1.31
8          11.1           1.09
9           7.5           1.18
10         12.2           1.22
```

11	6.7	1.25
12	5.2	1.19
13	19.0	1.95
14	15.1	1.28
15	6.7	1.52
16	8.6	NA
17	4.2	1.12
18	10.3	1.37
19	12.5	1.19
20	16.1	1.05
21	13.3	1.32
22	4.9	1.03
23	8.8	1.12
24	9.5	1.70

To read in factor variables (see Section 1.2.8), the easiest way may be to encode them using a textual representation. The `read.table` function autodetects whether a vector is text or numeric and converts it to a factor in the former case (but makes no attempt to recognize numerically coded factors). For instance, the `secretin` built-in data set is read from a file that begins like this:

```
   gluc person time repl time20plus time.comb
1    92      A  pre    a        pre      pre
2    93      A  pre    b        pre      pre
3    84      A   20    a        20+       20
4    88      A   20    b        20+       20
5    88      A   30    a        20+      30+
6    90      A   30    b        20+      30+
7    86      A   60    a        20+      30+
8    89      A   60    b        20+      30+
9    87      A   90    a        20+      30+
10   90      A   90    b        20+      30+
11   85      B  pre    a        pre      pre
12   85      B  pre    b        pre      pre
13   74      B   20    a        20+       20
....
```

This file can be read directly by `read.table` with no arguments other than the filename. It will recognize the case where the first line is one item shorter than the rest and will interpret that layout to imply that the first line contains a header and the first value on all subsequent lines is a row label — that is, exactly the layout generated when printing a data frame.

Reading factors like this may be convenient, but there is a drawback: The level order is alphabetic, so for instance

```
> levels(secretin$time)
[1] "20"   "30"   "60"   "90"   "pre"
```

If this is not what you want, then you may have to manipulate the factor levels; see Section 10.1.2.

A technical note: The files referenced above are contained in the ISwR package in the subdirectory (folder) rawdata. Exactly where the file is located on your system will depend on where the ISwR package was installed. You can find this out as follows:

```
> system.file("rawdata", "thuesen.txt", package="ISwR")
[1] "/home/pd/Rlibrary/ISwR/rawdata/thuesen.txt"
```

2.4.2 Further details on read.table

The read.table function is a very flexible tool that is controlled by many options. We shall not attempt a full description here but just give some indication of what it can do.

File format details

We have already seen the use of header=T. A couple of other options control the detailed format of the input file:

Field separator. This can be specified using sep. Notice that when this is used, as opposed to the default use of whitespace, there must be exactly one separator between data fields. Two consecutive separators will imply that there is a missing value in between. Conversely, it is necessary to use specific codes to represent missing values in the default format and also to use some form of quoting for strings that contain embedded spaces.

NA strings. You can specify which strings represent missing values via na.strings. There can be several different strings, although not different strings for different columns. For print files from the SAS program, you would use na.strings=".".

Quotes and comments. By default, R-style quotes can be used to delimit character strings, and parts of files following the comment character # are ignored. These features can be modified or removed via the quote and comment.char arguments.

Unequal field count. It is normally considered an error if not all lines contain the same number of values (the first line can be one item short, as described above for the secretin data). The fill and flush arguments can be used in case lines vary in length.

Delimited file types

Applications such as spreadsheets and databases produce text files in formats that require multiple options to be adjusted. For such purposes, there exist "precooked" variants of `read.table`. Two of these are intended to handle CSV files and are called `read.csv` and `read.csv2`. The former assumes that fields are separated by a comma, and the latter assumes that they are separated by semicolons but use a comma as the decimal point (this format is often generated in European locales). Both formats have `header=T` as the default. Further variants are `read.delim` and `read.delim2` for reading delimited files (by default, Tab-delimited files).

Conversion of input

It can be desirable to override the default conversion mechanisms in `read.table`. By default, nonnumeric input is converted to factors, but it does not always make sense. For instance, names and addresses typically should not be converted. This can be modified either for all columns using `stringsAsFactors` or on a per-item basis using `as.is`.

Automatic conversion is often convenient, but it is inefficient in terms of computer time and storage; in order to read a numeric column, `read.table` first reads it as character data, checks whether all elements can be converted to numeric, and only then performs the conversion. The `colClasses` argument allows you to bypass the mechanism by explicitly specifying which columns are of which class (the standard classes `"character"`, `"numeric"`, etc., get special treatment). You can also skip unwanted columns by specifying `"NULL"` as the class.

2.4.3 The data editor

R lets you edit data frames using a spreadsheet-like interface. The interface is a bit rough but quite useful for small data sets.

To edit a data frame, you can use the `edit` function:

```
> aq <- edit(airquality)
```

This brings up a spreadsheet-like editor with a column for each variable in the data frame. The `airquality` data set is built into R; see `help(airquality)` for its contents. Inside the editor, you can move around with the mouse or the cursor keys and edit the current cell by typing in data. The type of variable can be switched between real (numeric) and character (factor) by clicking on the column header, and the name of the variable can be changed similarly. Note that there is (as of R 2.6.2) no way to delete rows and columns and that new data can be entered only at the end.

When you close the data editor, the edited data frame is assigned to aq. The original airquality is left intact. Alternatively, if you do not mind overwriting the original data frame, you can use

```
> fix(aq)
```

This is equivalent to aq <- edit(aq).

To enter data into a blank data frame, use

```
> dd <- data.frame()
> fix(dd)
```

An alternative would be dd <- edit(data.frame()), which works fine except that beginners tend to reexecute the command when they need to edit dd, which of course destroys all data. It is necessary in either case to start with an empty data frame since by default edit expects you to want to edit a user-defined function and would bring up a text editor if you started it as edit().

2.4.4   Interfacing to other programs

Sometimes you will want to move data between R and other statistical packages or spreadsheets. A simple fallback approach is to request that the package in question export data as a text file of some sort and use read.table, read.csv, read.csv2, read.delim, or read.delim2, as previously described.

The foreign package is one of the packages labelled "recommended" and should therefore be available with binary distributions of R. It contains routines to read files in several formats, including those from SPSS (.sav format), SAS (export libraries), Epi-Info (.rec), Stata, Systat, Minitab, and some S-PLUS version 3 dump files.

Unix/Linux users sometimes find themselves with data sets written on Windows machines. The foreign package will work there as well for those formats that it supports. Notice that ordinary SAS data sets are not among the supported formats. These have to be converted to export libraries on the originating system. Data that have been entered into Microsoft Excel spreadsheets are most conveniently extracted using a compatible application such as OOo (OpenOffice.org).

An expedient technique is to read from the system clipboard. Say, highlight a rectangular region in a spreadsheet, press Ctrl-C (if on Windows), and inside R use

```
read.table("clipboard", header=T)
```

This does require a little caution, though. It may result in loss of accuracy since you only transfer the data as they appear on the screen. This is mostly a concern if you have data to many significant digits.

For data stored in databases, there exist a number of interface packages on CRAN. Of particular interest on Windows and with some Unix databases is the RODBC package because you can set up ODBC ("Open Database Connectivity") connections to data stored by common applications, including Excel and Access. Some Unix databases (e.g., PostgreSQL) also allow ODBC connections.

For up-to-date information on these matters, consult the "R Data Import/Export" manual that comes with the system.

2.5 Exercises

2.1 Describe how to insert a value between two elements of a vector at a given position by using the append function (use the help system to find out). Without append, how would you do it?

2.2 Write the built-in data set thuesen to a Tab-separated text file with write.table. View it with a text editor (depending on your system). Change the NA value to . (period), and read the changed file back into R with a suitable command. Also try importing the data into other applications of your choice and exporting them to a new file after editing. You may have to remove row names to make this work.

3

Probability and distributions

The concepts of randomness and probability are central to statistics. It is an empirical fact that most experiments and investigations are not perfectly reproducible. The degree of irreproducibility may vary: Some experiments in physics may yield data that are accurate to many decimal places, whereas data on biological systems are typically much less reliable. However, the view of data as something coming from a statistical distribution is vital to understanding statistical methods. In this section, we outline the basic ideas of probability and the functions that R has for random sampling and handling of theoretical distributions.

3.1 Random sampling

Much of the earliest work in probability theory was about games and gambling issues, based on symmetry considerations. The basic notion then is that of a random sample: dealing from a well-shuffled pack of cards or picking numbered balls from a well-stirred urn.

In R, you can simulate these situations with the `sample` function. If you want to pick five numbers at random from the set `1:40`, then you can write

```
> sample(1:40,5)
[1]  4 30 28 40 13
```

P. Dalgaard, *Introductory Statistics with R*,
DOI: 10.1007/978-0-387-79054-1_3, © Springer Science+Business Media, LLC 2008

The first argument (x) is a vector of values to be sampled and the second (size) is the sample size. Actually, sample(40,5) would suffice since a single number is interpreted to represent the length of a sequence of integers.

Notice that the default behaviour of sample is *sampling without replacement*. That is, the samples will not contain the same number twice, and size obviously cannot be bigger than the length of the vector to be sampled. If you want sampling with replacement, then you need to add the argument replace=TRUE.

Sampling with replacement is suitable for modelling coin tosses or throws of a die. So, for instance, to simulate 10 coin tosses we could write

```
> sample(c("H","T"), 10, replace=T)
 [1] "T" "T" "T" "T" "T" "H" "H" "T" "H" "T"
```

In fair coin-tossing, the probability of heads should equal the probability of tails, but the idea of a random event is not restricted to symmetric cases. It could be equally well applied to other cases, such as the successful outcome of a surgical procedure. Hopefully, there would be a better than 50% chance of this. You can simulate data with nonequal probabilities for the outcomes (say, a 90% chance of success) by using the prob argument to sample, as in

```
> sample(c("succ", "fail"), 10, replace=T, prob=c(0.9, 0.1))
 [1] "succ" "succ" "succ" "succ" "succ" "succ" "succ" "succ"
 [9] "succ" "succ"
```

This may not be the best way to generate such a sample, though. See the later discussion of the binomial distribution.

3.2   Probability calculations and combinatorics

Let us return to the case of sampling without replacement, specifically sample(1:40, 5). The probability of obtaining a given number as the first one of the sample should be 1/40, the next one 1/39, and so forth. The probability of a given sample should then be $1/(40 \times 39 \times 38 \times 37 \times 36)$. In R, use the prod function, which calculates the product of a vector of numbers

```
> 1/prod(40:36)
[1] 1.266449e-08
```

However, notice that this is the probability of getting given numbers in a given order. If this were a Lotto-like game, then you would rather be interested in the probability of guessing a given *set* of five numbers correctly. Thus you need also to include the cases that give the same numbers in a different order. Since obviously the probability of each such case is going to be the same, all we need to do is to figure out how many such cases there are and multiply by that. There are five possibilities for the first number, and for each of these there are four possibilities for the second, and so forth; that is, the number is $5 \times 4 \times 3 \times 2 \times 1$. This number is also written as 5! (5 *factorial*). So the probability of a "winning Lotto coupon" would be

```
> prod(5:1)/prod(40:36)
[1] 1.519738e-06
```

There is another way of arriving at the same result. Notice that since the actual set of numbers is immaterial, all sets of five numbers must have the same probability. So all we need to do is to calculate the number of ways to choose 5 numbers out of 40. This is denoted

$$\binom{40}{5} = \frac{40!}{5!35!} = 658008$$

In R, the choose function can be used to calculate this number, and the probability is thus

```
> 1/choose(40,5)
[1] 1.519738e-06
```

3.3 Discrete distributions

When looking at independent replications of a binary experiment, you would not usually be interested in whether each case is a success or a failure but rather in the total number of successes (or failures). Obviously, this number is random since it depends on the individual random outcomes, and it is consequently called a *random variable*. In this case it is a discrete-valued random variable that can take values $0, 1, \ldots, n$, where n is the number of replications. Continuous random variables are encountered later.

A random variable X has a *probability distribution* that can be described using *point probabilities* $f(x) = P(X = x)$ or the *cumulative distribution function* $F(x) = P(X \leq x)$. In the case at hand, the distribution can be worked out as having the point probabilities

$$f(x) = \binom{n}{x} p^x (1 - p)^{n-x}$$

This is known as the *binomial distribution*, and the $\binom{n}{x}$ are known as *binomial coefficients*. The parameter p is the probability of a successful outcome in an individual trial. A graph of the point probabilities of the binomial distribution appears in Figure 3.2 ahead.

We delay describing the R functions related to the binomial distribution until we have discussed continuous distributions so that we can present the conventions in a unified manner.

Many other distributions can be derived from simple probability models. For instance, the *geometric distribution* is similar to the binomial distribution but records the number of failures that occur before the first success.

3.4   Continuous distributions

Some data arise from measurements on an essentially continuous scale, for instance temperature, concentrations, etc. In practice, they will be recorded to a finite precision, but it is useful to disregard this in the modelling. Such measurements will usually have a component of random variation, which makes them less than perfectly reproducible. However, these random fluctuations will tend to follow patterns; typically they will cluster around a central value, with large deviations being more rare than smaller ones.

In order to model continuous data, we need to define random variables that can obtain the value of any real number. Because there are infinitely many numbers infinitely close, the probability of any particular value will be zero, so there is no such thing as a point probability as for discrete-valued random variables. Instead we have the concept of a *density*. This is the infinitesimal probability of hitting a small region around x divided by the size of the region. The cumulative distribution function can be defined as before, and we have the relation

$$F(x) = \int_{-\infty}^{x} f(x)\, dx$$

There are a number of standard distributions that come up in statistical theory and are available in R. It makes little sense to describe them in detail here except for a couple of examples.

The *uniform distribution* has a constant density over a specified interval (by default $[0, 1]$).

The *normal distribution* (also known as the *Gaussian distribution*) has density

$$f(x) = \frac{1}{\sqrt{2\pi}\sigma} \exp\left(-\frac{(x-\mu)^2}{2\sigma^2}\right)$$

depending on its mean μ and standard deviation σ. The normal distribution has a characteristic bell shape (Figure 3.1), and modifying μ and σ simply translates and widens the distribution. It is a standard building block in statistical models, where it is commonly used to describe error variation. It also comes up as an approximating distribution in several contexts; for instance, the binomial distribution for large sample sizes can be well approximated by a suitably scaled normal distribution.

3.5   The built-in distributions in R

The standard distributions that turn up in connection with model building and statistical tests have been built into R, and it can therefore completely replace traditional statistical tables. Here we look only at the normal distribution and the binomial distribution, but other distributions follow exactly the same pattern.

Four fundamental items can be calculated for a statistical distribution:

- Density or point probability
- Cumulated probability, distribution function
- Quantiles
- Pseudo-random numbers

For all distributions implemented in R, there is a function for each of the four items listed above. For example, for the normal distribution, these are named dnorm, pnorm, qnorm, and rnorm (density, probability, quantile, and random, respectively).

3.5.1   Densities

The density for a continuous distribution is a measure of the relative probability of "getting a value close to x". The probability of getting a value in a particular interval is the area under the corresponding part of the curve.

For discrete distributions, the term "density" is used for the point probability — the probability of getting exactly the value x. Technically, this is
correct: It is a density with respect to counting measure.

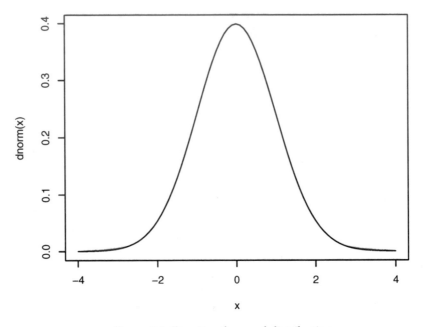

Figure 3.1. Density of normal distribution.

The density function is likely the one of the four function types that is least
used in practice, but if for instance it is desired to draw the well-known
bell curve of the normal distribution, then it can be done like this:

```
> x <- seq(-4,4,0.1)
> plot(x,dnorm(x),type="l")
```

(Notice that this is the letter 'l', not the digit '1').

The function seq (see p. 15) is used to generate equidistant values, here
from -4 to 4 in steps of 0.1; that is, $(-4.0, -3.9, -3.8, \ldots, 3.9, 4.0)$. The use
of type="l" as an argument to plot causes the function to draw lines
between the points rather than plotting the points themselves.

An alternative way of creating the plot is to use curve as follows:

```
> curve(dnorm(x), from=-4, to=4)
```

This is often a more convenient way of making graphs, but it does require that the *y*-values can be expressed as a simple functional expression in *x*.

For discrete distributions, where variables can take on only distinct values, it is preferable to draw a pin diagram, here for the binomial distribution with $n = 50$ and $p = 0.33$ (Figure 3.2):

```
> x <- 0:50
> plot(x,dbinom(x,size=50,prob=.33),type="h")
```

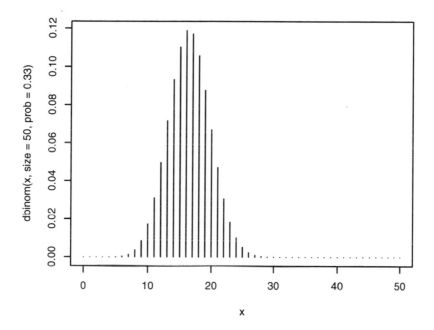

Figure 3.2. Point probabilities in binom(50, 0.33).

Notice that there are three arguments to the "d-function" this time. In addition to *x*, you have to specify the number of trials *n* and the probability parameter *p*. The distribution drawn corresponds to, for example, the number of 5s or 6s in 50 throws of a symmetrical die. Actually, `dnorm` also takes more than one argument, namely the mean and standard deviation, but they have default values of 0 and 1, respectively, since most often it is the standard normal distribution that is requested.

The form `0:50` is a short version of `seq(0,50,1)`: the whole numbers from 0 to 50 (see p. 15). It is `type="h"` (as in *h*istogram-like) that causes the pins to be drawn.

3.5.2   Cumulative distribution functions

The cumulative distribution function describes the probability of "hitting" *x* or less in a given distribution. The corresponding R functions begin with a 'p' (for probability) by convention.

Just as you can plot densities, you can of course also plot cumulative distribution functions, but that is usually not very informative. More often, actual numbers are desired. Say that it is known that some biochemical measure in healthy individuals is well described by a normal distribution with a mean of 132 and a standard deviation of 13. Then, if a patient has a value of 160, there is

```
> 1-pnorm(160,mean=132,sd=13)
[1] 0.01562612
```

or only about 1.5% of the general population, that has that value or higher. The function pnorm returns the probability of getting a value smaller than its first argument in a normal distribution with the given mean and standard deviation.

Another typical application occurs in connection with statistical tests. Consider a simple sign test: Twenty patients are given two treatments each (blindly and in randomized order) and then asked whether treatment A or B worked better. It turned out that 16 patients liked A better. The question is then whether this can be taken as sufficient evidence that A actually is the better treatment or whether the outcome might as well have happened by chance even if the treatments were equally good. If there was no difference between the two treatments, then we would expect the number of people favouring treatment A to be binomially distributed with $p = 0.5$ and $n = 20$. How (im)probable would it then be to obtain what we have observed? As in the normal distribution, we need a tail probability, and the immediate guess might be to look at

```
> pbinom(16,size=20,prob=.5)
[1] 0.9987116
```

and subtract it from 1 to get the upper tail — but this would be an error! What we need is the probability of *the observed or more extreme*, and pbinom is giving the probability of 16 or less. We need to use "15 or less" instead.

```
> 1-pbinom(15,size=20,prob=.5)
[1] 0.005908966
```

If you want a two-tailed test because you have no prior idea about which treatment is better, then you will have to add the probability of obtaining equally extreme results in the opposite direction. In the present case, that means the probability that four or fewer people prefer A, giving a total probability of

```
> 1-pbinom(15,20,.5)+pbinom(4,20,.5)
[1] 0.01181793
```

(which is obviously exactly twice the one-tailed probability).

As can be seen from the last command, it is not strictly necessary to use the `size` and `prob` keywords as long as the arguments are given in the right order (positional matching; see Section 1.2.2).

It is quite confusing to keep track of whether or not the observation itself needs to be counted. Fortunately, the function `binom.test` keeps track of such formalities and performs the correct binomial test. This is further discussed in Chapter 8.

3.5.3   Quantiles

The quantile function is the inverse of the cumulative distribution function. The p-quantile is the value with the property that there is probability p of getting a value less than or equal to it. The median is by definition the 50% quantile.

Some details concerning the definition in the case of discontinuous distributions are glossed over here. You can fairly easily deduce the behaviour by experimenting with the R functions.

Tables of statistical distributions are almost always given in terms of quantiles. For a fixed set of probabilities, the table shows the boundary that a test statistic must cross in order to be considered significant at that level. This is purely for operational reasons; it is almost superfluous when you have the option of computing p exactly.

Theoretical quantiles are commonly used for the calculation of confidence intervals and for power calculations in connection with designing and dimensioning experiments (see Chapter 9). A simple example of a confidence interval can be given here (see also Chapter 5).

If we have n normally distributed observations with the same mean μ and standard deviation σ, then it is known that the average $\bar{x}$ is normally distributed around μ with standard deviation $\sigma/\sqrt{n}$. A 95% confidence interval for μ can be obtained as

$$\bar{x} + \sigma/\sqrt{n} \times N_{0.025} \le \mu \le \bar{x} + \sigma/\sqrt{n} \times N_{0.975}$$

where $N_{0.025}$ is the 2.5% quantile in the normal distribution. If $\sigma = 12$ and we have measured $n = 5$ persons and found an average of $\bar{x} = 83$, then we can compute the relevant quantities as ("sem" means *standard error of the mean*)

```
> xbar <- 83
> sigma <- 12
> n <- 5
> sem <- sigma/sqrt(n)
> sem
[1] 5.366563
> xbar + sem * qnorm(0.025)
[1] 72.48173
> xbar + sem * qnorm(0.975)
[1] 93.51827
```

and thus find a 95% confidence interval for μ going from 72.48 to 93.52. (Notice that this is based on the assumption that σ is known. This is sometimes reasonable in process control applications. The more common case of estimating σ from the data leads to confidence intervals based on the t distribution and is discussed in Chapter 5.)

Since it is known that the normal distribution is symmetric, so that $N_{0.025} = -N_{0.975}$, it is common to write the formula for the confidence interval as $\bar{x} \pm \sigma/\sqrt{n} \times N_{0.975}$. The quantile itself is often written $\Phi^{-1}(0.975)$, where Φ is standard notation for the cumulative distribution function of the normal distribution (pnorm).

Another application of quantiles is in connection with Q–Q plots (see Section 4.2.3), which can be used to assess whether a set of data can reasonably be assumed to come from a given distribution.

3.5.4   Random numbers

To many people, it sounds like a contradiction in terms to generate random numbers on a computer since its results are supposed to be predictable and reproducible. What is in fact possible is to generate sequences of "pseudo-random" numbers, which for practical purposes behave *as if* they were drawn randomly.

Here random numbers are used to give the reader a feeling for the way in which randomness affects the quantities that can be calculated from a set of data. In professional statistics, they are used to create simulated data sets in order to study the accuracy of mathematical approximations and the effect of assumptions being violated.

The use of the functions that generate random numbers is straightforward. The first argument specifies the number of random numbers to compute, and the subsequent arguments are similar to those for other functions related to the same distributions. For instance,

```
> rnorm(10)
 [1] -0.2996466 -0.1718510 -0.1955634  1.2280843 -2.6074190
 [6] -0.2999453 -0.4655102 -1.5680666  1.2545876 -1.8028839
> rnorm(10)
 [1]  1.7082495  0.1432875 -1.0271750 -0.9246647  0.6402383
 [6]  0.7201677 -0.3071239  1.2090712  0.8699669  0.5882753
> rnorm(10,mean=7,sd=5)
 [1]  8.934983  8.611855  4.675578  3.670129  4.223117  5.484290
 [7] 12.141946  8.057541 -2.893164 13.590586
> rbinom(10,size=20,prob=.5)
 [1] 12 11 10  8 11  8 11  8  8 13
```

3.6 Exercises

3.1 Calculate the probability for each of the following events: (a) A standard normally distributed variable is larger than 3. (b) A normally distributed variable with mean 35 and standard deviation 6 is larger than 42. (c) Getting 10 out of 10 successes in a binomial distribution with probability 0.8. (d) $X < 0.9$ when X has the standard uniform distribution. (e) $X > 6.5$ in a χ^2 distribution with 2 degrees of freedom.

3.2 A rule of thumb is that 5% of the normal distribution lies outside an interval approximately $\pm 2s$ about the mean. To what extent is this true? Where are the limits corresponding to 1%, 0.5%, and 0.1%? What is the position of the quartiles measured in standard deviation units?

3.3 For a disease known to have a postoperative complication frequency of 20%, a surgeon suggests a new procedure. He tests it on 10 patients and there are no complications. What is the probability of operating on 10 patients successfully with the traditional method?

3.4 Simulated coin-tossing can be done using rbinom instead of sample. How exactly would you do that?

4

Descriptive statistics and graphics

Before going into the actual statistical modelling and analysis of a data set, it is often useful to make some simple characterizations of the data in terms of summary statistics and graphics.

4.1 Summary statistics for a single group

It is easy to calculate simple summary statistics with R. Here is how to calculate the mean, standard deviation, variance, and median.

```
> x <- rnorm(50)
> mean(x)
[1] 0.03301363
> sd(x)
[1] 1.069454
> var(x)
[1] 1.143731
> median(x)
[1] -0.08682795
```

[handwritten annotation: random numbers produce diff results]

Notice that the example starts with the generation of an artificial data vector x of 50 normally distributed observations. It is used in examples throughout this section. When reproducing the examples, you will not get exactly the same results since your random numbers will differ.

P. Dalgaard, *Introductory Statistics with R*,
DOI: 10.1007/978-0-387-79054-1_4, © Springer Science+Business Media, LLC 2008

Empirical quantiles may be obtained with the function `quantile` like this:

3 quartiles (handwritten)

```
> quantile(x)
        0%          25%          50%          75%         100%
-2.60741896  -0.54495849  -0.08682795   0.70018536   2.98872414
```

As you see, by default you get the minimum, the maximum, and the three *quartiles* — the 0.25, 0.50, and 0.75 quantiles — so named because they correspond to a division into four parts. Similarly, we have *deciles* for $0.1, 0.2, \ldots, 0.9$, and *centiles* or *percentiles*. The difference between the first and third quartiles is called the *interquartile range* (IQR) and is sometimes used as a robust alternative to the standard deviation.

It is also possible to obtain other quantiles; this is done by adding an argument containing the desired percentage points. This, for example, is how to get the deciles:

$0 \rightarrow 1$ by 0.1 (handwritten)

```
> pvec <- seq(0,1,0.1)
> pvec
 [1] 0.0 0.1 0.2 0.3 0.4 0.5 0.6 0.7 0.8 0.9 1.0
> quantile(x,pvec)
         0%          10%          20%          30%          40%
-2.60741896  -1.07746896  -0.70409272  -0.46507213  -0.29976610
        50%          60%          70%          80%          90%
-0.08682795   0.19436950   0.49060129   0.90165137   1.31873981
       100%
 2.98872414
```

Be aware that there are several possible definitions of empirical quantiles. The one R uses by default is based on a sum polygon where the ith ranking observation is the $(i-1)/(n-1)$ quantile and intermediate quantiles are obtained by linear interpolation. It sometimes confuses students that in a sample of 10 there will be 3 observations below the first quartile with this definition. Other definitions are available via the `type` argument to `quantile`.

If there are missing values in data, things become a bit more complicated. For illustration, we use the following example.

The data set `juul` contains variables from an investigation performed by Anders Juul (Rigshospitalet, Department for Growth and Reproduction) concerning serum IGF-I (insulin-like growth factor) in a group of healthy humans, primarily schoolchildren. The data set is contained in the `ISwR` package and contains a number of variables, of which we only use `igf1` (serum IGF-I) for now, but later in the chapter we also use `tanner` (Tanner stage of puberty, a classification into five groups based on appearance of primary and secondary sexual characteristics), sex, and menarche (indicating whether or not a girl has had her first period).

Attempting to calculate the mean of igf1 reveals a problem.

```
> attach(juul)
> mean(igf1)
[1] NA
```

R will not skip missing values unless explicitly requested to do so. The mean of a vector with an unknown value is unknown. However, you can give the na.rm argument (*not available, remove*) to request that missing values be removed:

```
> mean(igf1,na.rm=T)       na.rm
[1] 340.168
```

There is one slightly annoying exception: The length function will not understand na.rm, so we cannot use it to count the number of nonmissing measurements of igf1. However, you can use

```
> sum(!is.na(igf1))
[1] 1018
```

The construction above uses the fact that if logical values are used in arithmetic, then TRUE is converted to 1 and FALSE to 0.

A nice summary display of a numeric variable is obtained from the summary function:

```
> summary(igf1)
   Min. 1st Qu.  Median    Mean 3rd Qu.    Max.   NA's
   25.0   202.2   313.5   340.2   462.8   915.0   321.0
```

The 1st Qu. and 3rd Qu. refer to the empirical quartiles (0.25 and 0.75 quantiles).

In fact, it is possible to summarize an entire data frame with

```
> summary(juul)
      age             menarche              sex
 Min.   : 0.170   Min.   : 1.000   Min.   :1.000
 1st Qu.: 9.053   1st Qu.: 1.000   1st Qu.:1.000
 Median :12.560   Median : 1.000   Median :2.000
 Mean   :15.095   Mean   : 1.476   Mean   :1.534
 3rd Qu.:16.855   3rd Qu.: 2.000   3rd Qu.:2.000
 Max.   :83.000   Max.   : 2.000   Max.   :2.000
 NA's   : 5.000   NA's   :635.000  NA's   :5.000
      igf1             tanner             testvol
 Min.   : 25.0   Min.   : 1.000   Min.   : 1.000
 1st Qu.:202.2   1st Qu.: 1.000   1st Qu.: 1.000
```

```
Median :313.5    Median :   2.000   Median :   3.000
Mean   :340.2    Mean   :   2.640   Mean   :   7.896
3rd Qu.:462.8    3rd Qu.:   5.000   3rd Qu.:  15.000
Max.   :915.0    Max.   :   5.000   Max.   :  30.000
NA's   :321.0    NA's   :240.000    NA's   :859.000
```

The data set has `menarche`, `sex`, and `tanner` coded as numeric variables even though they are clearly categorical. This can be mended as follows:

```
> detach(juul)
> juul$sex <- factor(juul$sex,labels=c("M","F"))
> juul$menarche <- factor(juul$menarche,labels=c("No","Yes"))
> juul$tanner <- factor(juul$tanner,
+                       labels=c("I","II","III","IV","V"))
> attach(juul)
> summary(juul)
       age          menarche      sex           igf1
 Min.   : 0.170   No  :369    M   :621    Min.   :  25.0
 1st Qu.: 9.053   Yes :335    F   :713    1st Qu.: 202.2
 Median :12.560   NA's:635    NA's:  5    Median : 313.5
 Mean   :15.095                           Mean   : 340.2
 3rd Qu.:16.855                           3rd Qu.: 462.8
 Max.   :83.000                           Max.   : 915.0
 NA's   : 5.000                           NA's   : 321.0
 tanner         testvol
 I   :515    Min.   :  1.000
 II  :103    1st Qu.:  1.000
 III : 72    Median :  3.000
 IV  : 81    Mean   :  7.896
 V   :328    3rd Qu.: 15.000
 NA's:240    Max.   : 30.000
             NA's   :859.000
```

Notice how the display changes for the factor variables. Note also that `juul` was detached and reattached after the modification. This is because modifying a data frame does not affect any attached version. It was not strictly necessary to do it here because `summary` works directly on the data frame whether attached or not.

In the above, the variables `sex`, `menarche`, and `tanner` were converted to factors with suitable level names (in the raw data these are represented using numeric codes). The converted variables were put back into the data frame `juul`, replacing the original `sex`, `tanner`, and `menarche` variables. We might also have used the `transform` function (or `within`):

```
> juul <- transform(juul,
+    sex=factor(sex,labels=c("M","F")),
+    menarche=factor(menarche,labels=c("No","Yes")),
+    tanner=factor(tanner,labels=c("I","II","III","IV","V")))
```

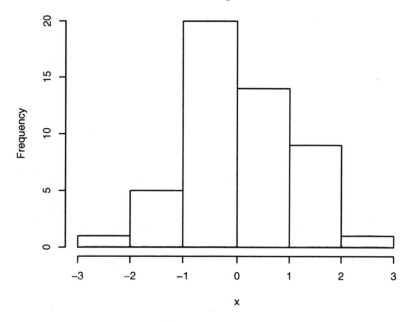

Figure 4.1. Histogram.

4.2 Graphical display of distributions

4.2.1 Histograms

You can get a reasonable impression of the shape of a distribution by drawing a histogram; that is, a count of how many observations fall within specified divisions ("bins") of the x-axis (Figure 4.1).

```
> hist(x)
```

By specifying breaks=n in the hist call, you get *approximately n* bars in the histogram since the algorithm tries to create "pretty" cutpoints. You can have full control over the interval divisions by specifying breaks as a vector rather than as a number. Altman (1991, pp. 25–26) contains an example of accident rates by age group. These are given as a count in age groups 0–4, 5–9, 10–15, 16, 17, 18–19, 20–24, 25–59, and 60–79 years of age. The data can be entered as follows:

```
> mid.age <- c(2.5,7.5,13,16.5,17.5,19,22.5,44.5,70.5)
> acc.count <- c(28,46,58,20,31,64,149,316,103)
> age.acc <- rep(mid.age,acc.count)
```

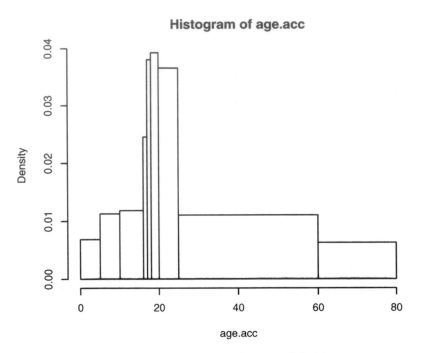

Figure 4.2. Histogram with unequal divisions.

```
> brk <- c(0,5,10,16,17,18,20,25,60,80)
> hist(age.acc,breaks=brk)
```

Here the first three lines generate pseudo-data from the table in the book. For each interval, the relevant number of "observations" is generated with an age set to the midpoint of the interval; that is, 28 2.5-year-olds, 46 7.5-year-olds, etc. Then a vector `brk` of cutpoints is defined (note that the extremes need to be included) and used as the `breaks` argument to `hist`, yielding Figure 4.2.

Notice that you automatically got the "correct" histogram where the _area of a column is proportional to the number._ The y-axis is in density units (that is, proportion of data per x unit), so that the total area of the histogram will be 1. If, for some reason, you want the (misleading) histogram where the column height is the raw number in each interval, then it can be specified using `freq=T`. For equidistant breakpoints, that is the default (because then you can see how many observations have gone into each column), but you can set `freq=F` to get densities displayed. This is really just a change of scale on the y-axis, but it has the advantage that it becomes possible to overlay the histogram with a corresponding theoretical density function.

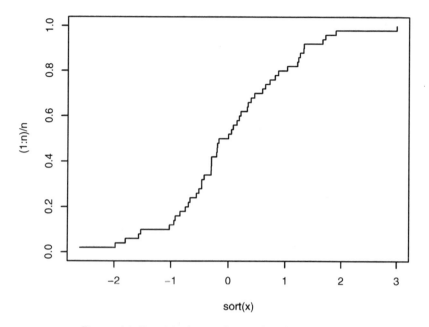

Figure 4.3. Empirical cumulative distribution function.

4.2.2 Empirical cumulative distribution

The empirical cumulative distribution function is defined as the fraction of data smaller than or equal to x. That is, if x is the kth smallest observation, then the proportion k/n of the data is smaller than or equal to x (7/10 if x is no. 7 of 10). The empirical cumulative distribution function can be plotted as follows (see Figure 4.3) where x is the simulated data vector from Section 4.1:

```
> n <- length(x)
> plot(sort(x),(1:n)/n,type="s",ylim=c(0,1))
```

The plotting parameter type="s" gives a step function where (x, y) is the left end of the steps and ylim is a vector of two elements specifying the extremes of the y-coordinates on the plot. Recall that c(...) is used to create vectors.

Some more elaborate displays of empirical cumulative distribution functions are available via the ecdf function. This is also more precise regarding the mathematical definition of the step function.

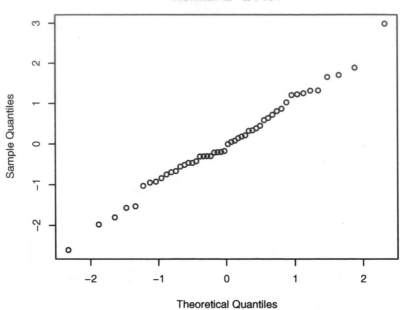

Figure 4.4. Q–Q plot using `qqnorm(x)`.

4.2.3 Q–Q plots

One purpose of calculating the empirical cumulative distribution function (c.d.f.) is to see whether data can be assumed normally distributed. For a better assessment, you might plot the kth smallest observation against the expected value of the kth smallest observation out of n in a standard normal distribution. The point is that in this way you would expect to obtain a straight line if data come from a normal distribution with *any* mean and standard deviation.

Creating such a plot is slightly complicated. Fortunately, there is a built-in function for doing it, `qqnorm`. The result of using it can be seen in Figure 4.4. You only have to write

```
> qqnorm(x)
```

As the title of the plot indicates, plots of this kind are also called "Q–Q plots" (quantile versus quantile). Notice that x and y are interchanged relative to the empirical c.d.f. — the observed values are now drawn along the y-axis. You should notice that with this convention the distribution has heavy tails if the outer parts of the curve are steeper than the middle part.

Some readers will have been taught "probability plots", which are similar but have the axes interchanged. It can be argued that the way R draws the plot is the better one since the theoretical quantiles are known in advance, while the empirical quantiles depend on data. You would normally choose to draw fixed values horizontally and variable values vertically.

4.2.4 Boxplots

A "boxplot", or more descriptively a "box-and-whiskers plot", is a graphical summary of a distribution. Figure 4.5 shows boxplots for IgM and its logarithm; see the example on page 23 in Altman (1991).

Here is how a boxplot is drawn in R. The box in the middle indicates "hinges" (nearly quartiles; see the help page for boxplot.stats) and median. The lines ("whiskers") show the largest or smallest observation that falls within a distance of 1.5 times the box size from the nearest hinge. If any observations fall farther away, the additional points are considered "extreme" values and are shown separately.

The practicalities are these:

```
> par(mfrow=c(1,2))
> boxplot(IgM)
> boxplot(log(IgM))
> par(mfrow=c(1,1))
```

A layout with two plots side by side is specified using the mfrow graphical parameter. It should be read as "multiframe, rowwise, 1 × 2 layout". Individual plots are organized in one row and two columns. As you might guess, there is also an mfcol parameter to plot columnwise. In a 2 × 2 layout, the difference is whether plot no. 2 is drawn in the top right or bottom left corner.

Notice that it is necessary to reset the layout parameter to c(1,1) at the end unless you also want two plots side by side subsequently.

4.3 Summary statistics by groups

When dealing with grouped data, you will often want to have various summary statistics computed within groups; for example, a table of means and standard deviations. To this end, you can use tapply (see Section 1.2.15). Here is an example concerning the folate concentration in red blood cells according to three types of ventilation during anesthesia (Alt-

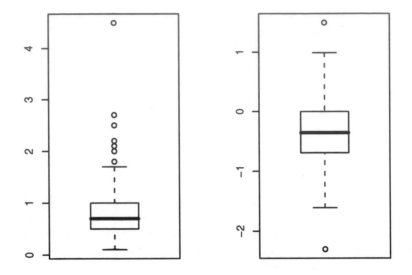

Figure 4.5. Boxplots for IgM and log IgM.

man, 1991, p. 208). We return to this example in Section 7.1, which also contains the explanation of the category names.

```
> attach(red.cell.folate)
> tapply(folate,ventilation,mean)
N2O+O2,24h  N2O+O2,op      O2,24h
  316.6250    256.4444    278.0000
```

The `tapply` call takes the `folate` variable, splits it according to `ventilation`, and computes the mean for each group. In the same way, standard deviations and the number of observations in the groups can be computed.

```
> tapply(folate,ventilation,sd)
N2O+O2,24h  N2O+O2,op      O2,24h
  58.71709    37.12180    33.75648
> tapply(folate,ventilation,length)
N2O+O2,24h  N2O+O2,op      O2,24h
         8          9           5
```

Try something like this for a nicer display:

```
> xbar <- tapply(folate, ventilation, mean)
> s <- tapply(folate, ventilation, sd)
> n <- tapply(folate, ventilation, length)
> cbind(mean=xbar, std.dev=s, n=n)
                mean    std.dev n
N2O+O2,24h  316.6250 58.71709 8
N2O+O2,op   256.4444 37.12180 9
O2,24h      278.0000 33.75648 5
```

For the `juul` data, we might want the mean `igf1` by `tanner` group, but of course we run into the problem of missing values again:

```
> tapply(igf1, tanner, mean)
  I   II III   IV    V
 NA   NA  NA   NA   NA
```

We need to get `tapply` to pass `na.rm=T` as a parameter to `mean` to make it exclude the missing values. This is achieved simply by passing it as an additional argument to `tapply`.

```
> tapply(igf1, tanner, mean, na.rm=T)
         I        II       III        IV         V
207.4727 352.6714 483.2222 513.0172 465.3344
```

The functions `aggregate` and `by` are variations on the same topic. The former is very much like `tapply`, except that it works on an entire data frame and presents its results as a data frame. This is useful for presenting many variables at once; e.g.,

```
> aggregate(juul[c("age","igf1")],
+           list(sex=juul$sex), mean, na.rm=T)
  sex      age      igf1
1   M 15.38436 310.8866
2   F 14.84363 368.1006
```

Notice that the grouping argument in this case must be a list, even when it is one-dimensional, and that the names of the list elements get used as column names in the output. Notice also that since the function is applied to all columns of the data frame, you may have to choose a subset of columns, in this case the numeric variables.

The indexing variable is not necessarily part of the data frame that is being aggregated, and there is no attempt at "smart evaluation" as there is in `subset`, so you have to spell out `juul$sex`. You can also use the fact that data frames are list-like and say

```
> aggregate(juul[c("age","igf1")], juul["sex"], mean, na.rm=T)
  sex      age      igf1
1   M 15.38436 310.8866
2   F 14.84363 368.1006
```

(the "trick" being that indexing a data frame with single brackets yields a data frame as the result).

The by function is again similar, but different. The difference is that the function now takes an entire (sub-) data frame as its argument, so that you can for instance summarize the Juul data by sex as follows:

```
> by(juul, juul["sex"], summary)
sex: M
      age              menarche    sex          igf1            tanner
 Min.    : 0.17   No   :  0   M:621    Min.    : 29.0    I    :291
 1st Qu.: 8.85    Yes  :  0   F:  0    1st Qu.:176.0    II   : 55
 Median :12.38    NA's:621             Median :280.0    III  : 34
 Mean    :15.38                        Mean    :310.9    IV   : 41
 3rd Qu.:16.77                         3rd Qu.:430.2    V    :124
 Max.    :83.00                        Max.    :915.0    NA's: 76
                                       NA's    :145.0
      testvol
 Min.    :  1.000
 1st Qu.:  1.000
 Median :  3.000
 Mean    :  7.896
 3rd Qu.: 15.000
 Max.    : 30.000
 NA's    :141.000
-------------------------------------------------
sex: F
      age              menarche    sex          igf1            tanner
 Min.    : 0.25   No   :369   M:  0    Min.    : 25.0    I    :224
 1st Qu.: 9.30    Yes  :335   F:713    1st Qu.:233.0    II   : 48
 Median :12.80    NA's:  9             Median :352.0    III  : 38
 Mean    :14.84                        Mean    :368.1    IV   : 40
 3rd Qu.:16.93                         3rd Qu.:483.0    V    :204
 Max.    :75.12                        Max.    :914.0    NA's:159
                                       NA's    :176.0
      testvol
 Min.    : NA
 1st Qu.: NA
 Median : NA
 Mean    :NaN
 3rd Qu.: NA
 Max.    : NA
 NA's    :713
```

The result of the call to by is actually a list of objects that has has been wrapped as an object of class "by" and printed using a print method for that class. You can assign the result to a variable and access the result for each subgroup using standard list indexing.

The same technique can also be used to generate more elaborate statistical analyses for each group.

4.4 Graphics for grouped data

In dealing with grouped data, it is important to be able not only to create plots for each group but also to compare the plots between groups. In this section we review some general graphical techniques that allow us to display similar plots for several groups on the same page. Some functions have specific features for displaying data from more than one group.

4.4.1 Histograms

We have already seen in Section 4.2.1 how to obtain a histogram simply by typing `hist(x)`, where x is the variable containing the data. R will then choose a number of groups so that a reasonable number of data points fall in each bin while at the same time ensuring that the cutpoints are "pretty" numbers on the x-axis.

It is also mentioned there that an alternative number of intervals can be set via the argument `breaks`, although you do not always get exactly the number you asked for since R reserves the right to choose "pretty" column boundaries. For instance, multiples of 0.5 MJ are chosen in the following example using the `energy` data introduced in Section 1.2.14 on the 24-hour energy expenditure for two groups of women.

In this example, some further techniques of general use are illustrated. The end result is seen in Figure 4.6, but first we must fetch the data:

```
> attach(energy)
> expend.lean <- expend[stature=="lean"]
> expend.obese <- expend[stature=="obese"]
```

Notice how we separate the `expend` vector in the `energy` data frame into two vectors according to the value of the factor `stature`.

Now we do the actual plotting:

```
> par(mfrow=c(2,1))
> hist(expend.lean,breaks=10,xlim=c(5,13),ylim=c(0,4),col="white")
> hist(expend.obese,breaks=10,xlim=c(5,13),ylim=c(0,4),col="grey")
> par(mfrow=c(1,1))
```

We set `par(mfrow=c(2,1))` to get the two histograms in the same plot. In the `hist` commands themselves, we used the `breaks` argument as already mentioned and `col`, whose effect should be rather obvious. We also used `xlim` and `ylim` to get the same x and y axes in the two plots. However, it is a coincidence that the columns have the same width.

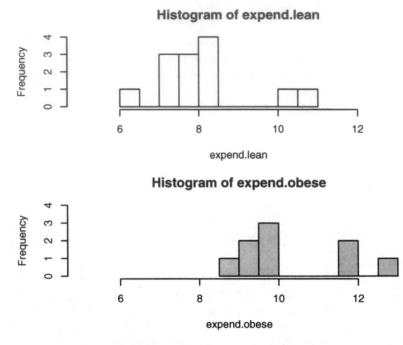

Figure 4.6. Histograms with refinements.

As a practical remark, when working with plots like the above, where more than a single line of code is required, it gets cumbersome to use command recall in the R console window every time something needs to be changed. A better idea may be to start up a script window or a plain-text editor and cut and paste entire blocks of code from there (see Section 2.1.3). You might also take it as an incentive to start writing simple functions.

4.4.2 Parallel boxplots

You might want a set of boxplots from several groups in the same frame. `boxplot` can handle this both when data are given in the form of separate vectors from each group and when data are in one long vector and a parallel vector or factor defines the grouping. To illustrate the latter, we use the `energy` data introduced in Section 1.2.14.

Figure 4.7 is created as follows:

```
> boxplot(expend ~ stature)
```

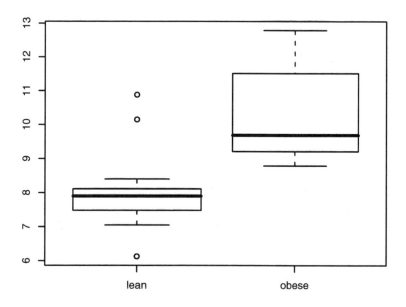

Figure 4.7. Parallel boxplot.

We could also have based the plot on the separate vectors `expend.lean` and `expend.obese`. In that case, a syntax is used that specifies the vectors as two separate arguments:

```
> boxplot(expend.lean,expend.obese)
```

The plot is not shown here, but the only difference lies in the labelling of the *x*-axis. There is also a third form, where data are given as a single argument that is a list of vectors.

The bottom plot has been made using the complete `expend` vector and the grouping variable `fstature`.

Notation of the type `y ~ x` should be read "y described using x". This is the first example we see of a *model formula*. We see many more examples of model formulas later on.

4.4.3 Stripcharts

The boxplots made in the preceding section show a "Laurel & Hardy" effect that is not really well founded in the data. The cause is that the in-

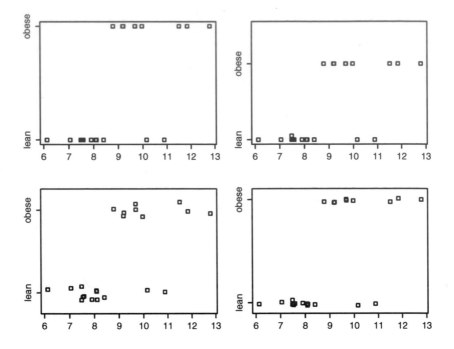

Figure 4.8. Stripcharts in four variations.

terquartile range is quite a bit larger in one group than in the other, making
the boxplot appear "fatter". With groups as small as these, the quartiles
will be quite inaccurately determined, and it may therefore be more desirable to plot the raw data. If you were to do this by hand, you might draw a
dot diagram where every number is marked with a dot on a number line.
R's automated variant of this is the function `stripchart`. Four variants
of stripcharts are shown in Figure 4.8.

The four plots were created as follows:

```
> opar <- par(mfrow=c(2,2), mex=0.8, mar=c(3,3,2,1)+.1)
> stripchart(expend ~ stature)
> stripchart(expend ~ stature, method="stack")
> stripchart(expend ~ stature, method="jitter")
> stripchart(expend ~ stature, method="jitter", jitter=.03)
> par(opar)
```

Notice that a little `par` magic was used to reduce the spacing between the
four plots. The `mex` setting reduces the interline distance, and `mar` reduces
the number of lines that surround the plot region. This can be done for
these plots since they have neither main title, subtitle, nor x and y labels.

All the original values of the changed settings can be stored in a variable (here `opar`) and reestablished with `par(opar)`.

The first plot is a standard stripchart, where the points are simply plotted on a line. The problem with this is that some points can become invisible because they are overplotted. This is why there is a `method` argument, which can be set to either `"stack"` or `"jitter"`.

The former method stacks points with identical values, but it only does so if data are *completely identical,* so in the upper right plot, it is only the two replicates of 7.48 that get stacked, whereas 8.08, 8.09, and 8.11 are still plotted in almost the same spot.

The "jitter" method offsets all points a random amount vertically. The standard jittering on plot no. 3 (bottom left) is a bit large; it may be preferable to make it clearer that data are placed along a horizontal line. For that purpose, you can set `jitter` lower than the default of 0.1, which is done in the fourth plot.

In this example we have not bothered to specify data in several forms as we did for `boxplot` but used `expend~stature` throughout. We could also have written

```
stripchart(list(lean=expend.lean, obese=expend.obese))
```

but `stripchart(expend.lean, expend.obese)` cannot be used.

4.5   Tables

Categorical data are usually described in the form of tables. This section outlines how you can create tables from your data and calculate relative frequencies.

4.5.1   Generating tables

We deal mainly with two-way tables. In the first example, we enter a table directly, as is required for tables taken from a book or a journal article.

A two-way table can be entered as a matrix object (Section 1.2.7). Altman (1991, p. 242) contains an example on caffeine consumption by marital status among women giving birth. That table may be input as follows:

```
> caff.marital <- matrix(c(652,1537,598,242,36,46,38,21,218
+ ,327,106,67),
+ nrow=3,byrow=T)
> caff.marital
     [,1] [,2] [,3] [,4]
[1,]  652 1537  598  242
[2,]   36   46   38   21
[3,]  218  327  106   67
```

The matrix function needs an argument containing the table values as a single vector and also the number of rows in the argument nrow. By default, the values are entered columnwise; if rowwise entry is desired, then you need to specify byrow=T.

You might also give the number of columns instead of rows using ncol. If exactly one of ncol and nrow is given, R will compute the other one so that it fits the number of values. If both ncol and nrow are given and it does not fit the number of values, the values will be "recycled", which in some (other!) circumstances can be useful.

To get readable printouts, you can add row and column names to the matrices.

```
> colnames(caff.marital) <- c("0","1-150","151-300",">300")
> rownames(caff.marital) <- c("Married","Prev.married","Single")
> caff.marital
               0 1-150 151-300 >300
Married      652  1537     598  242
Prev.married  36    46      38   21
Single       218   327     106   67
```

Furthermore, you can name the row and column names as follows. This is particularly useful if you are generating many tables with similar classification criteria.

```
> names(dimnames(caff.marital)) <- c("marital","consumption")
> caff.marital
               consumption
marital          0 1-150 151-300 >300
  Married      652  1537     598  242
  Prev.married  36    46      38   21
  Single       218   327     106   67
```

Actually, I glossed over something. Tables are not completely equivalent to matrices. There is a "table" class for which special methods exist, and you can convert to that class using as.table(caff.marital). The table function below returns an object of class "table".

For most elementary purposes, you can use matrices where two-dimensional tables are expected. One important case where you do need `as.table` is when converting a table to a data frame of counts:

```
> as.data.frame(as.table(caff.marital))
           marital consumption Freq
1          Married           0  652
2     Prev.married           0   36
3           Single           0  218
4          Married       1-150 1537
5     Prev.married       1-150   46
6           Single       1-150  327
7          Married     151-300  598
8     Prev.married     151-300   38
9           Single     151-300  106
10         Married        >300  242
11    Prev.married        >300   21
12          Single        >300   67
```

In practice, the more frequent case is that you have a data frame with variables for each person in a data set. In that case, you should do the tabulation with `table`, `xtabs`, or `ftable`. These functions will generally work for tabulating numeric vectors as well as factor variables, but the latter will have their levels used for row and column names automatically. Hence, it is recommended to convert numerically coded categorical data into factors. The `table` function is the oldest and most basic of the three. The two others offer formula-based interfaces and better printing of multiway tables.

The data set `juul` was introduced on p. 68. Here we look at some other variables in that data set, namely `sex` and `menarche`; the latter indicates whether or not a girl has had her first period. We can generate some simple tables as follows:

```
> table(sex)
sex
  M   F
621 713
> table(sex,menarche)
   menarche
sex  No Yes
  M   0   0
  F 369 335
> table(menarche,tanner)
        tanner
menarche   I II III IV   V
     No  221 43  32 14   2
     Yes   1  1   5 26 202
```

Of course, the table of menarche versus sex is just a check on internal consistency of the data. The table of menarche versus Tanner stage of puberty is more interesting.

There are also tables with more than two sides, but not many simple statistical functions use them. Briefly, to tabulate such data, just write, for example, table(factor1, factor2, factor3). To input a table of cell counts, use the array function (an analogue of matrix).

The xtabs function is quite similar to table except that it uses a model formula interface. This most often uses a one-sided formula where you just list the classification variables separated by +.

```
> xtabs(~ tanner + sex, data=juul)
       sex
tanner   M    F
    I    291  224
    II    55   48
    III   34   38
    IV    41   40
    V    124  204
```

Notice how the interface allows you to refer to variables in a data frame without attaching it. The empty left-hand side can be replaced by a vector of counts in order to handle pretabulated data.

The formatting of multiway tables from table or xtabs is not really nice; e.g.,

```
> xtabs(~ dgn + diab + coma, data=stroke)
, , coma = No

       diab
dgn      No  Yes
   ICH   53    6
   ID   143   21
   INF  411   64
   SAH   38    0

, , coma = Yes

       diab
dgn      No  Yes
   ICH   19    1
   ID    23    3
   INF   23    2
   SAH    9    0
```

As you add dimensions, you get more of these two-sided subtables and it becomes rather easy to lose track. This is where ftable comes in. This function creates "flat" tables; e.g., like this:

```
> ftable(coma + diab ~ dgn, data=stroke)
     coma   No       Yes
     diab   No Yes   No Yes
dgn
ICH          53    6   19    1
ID          143   21   23    3
INF         411   64   23    2
SAH          38    0    9    0
```

That is, variables on the left-hand side tabulate across the page and those on the right tabulate downwards. ftable works on raw data as shown, but its data argument can also be a table as generated by one of the other functions.

Like any matrix, a table can be transposed with the t function:

```
> t(caff.marital)
             marital
consumption Married Prev.married Single
      0         652           36    218
   1-150       1537           46    327
 151-300        598           38    106
   >300         242           21     67
```

For multiway tables, exchanging indices (generalized transposition) is done by aperm.

4.5.2 Marginal tables and relative frequency

It is often desired to compute marginal tables; that is, the sums of the counts along one or the other dimension of a table. Due to missing values, this might not coincide with just tabulating a single factor. This is done fairly easily using the apply function (Section 1.2.15), but there is also a simplified version called margin.table, described below.

First, we need to generate the table itself:

```
> tanner.sex <- table(tanner,sex)
```

(tanner.sex is an arbitrarily chosen name for the crosstable.)

```
> tanner.sex
       sex
tanner   M    F
    I   291  224
   II    55   48
  III    34   38
   IV    41   40
    V   124  204
```

Then we compute the marginal tables:

```
> margin.table(tanner.sex,1)
tanner
  I  II III  IV   V
515 103  72  81 328
> margin.table(tanner.sex,2)
sex
  M   F
545 554
```

The second argument to margin.table is the number of the marginal index: 1 and 2 give row and column totals, respectively.

Relative frequencies in a table are generally expressed as proportions of the row or column totals. Tables of relative frequencies can be constructed using prop.table as follows:

```
> prop.table(tanner.sex,1)
      sex
tanner         M         F
   I   0.5650485 0.4349515
   II  0.5339806 0.4660194
   III 0.4722222 0.5277778
   IV  0.5061728 0.4938272
   V   0.3780488 0.6219512
```

Note that the *rows* (1st index) sum to 1. If a table of percentages is desired, just multiply the entire table by 100.

prop.table cannot be used to express the numbers relative to the grand total of the table, but you can of course always write

```
> tanner.sex/sum(tanner.sex)
      sex
tanner          M          F
   I   0.26478617 0.20382166
   II  0.05004550 0.04367607
   III 0.03093722 0.03457689
   IV  0.03730664 0.03639672
   V   0.11282985 0.18562329
```

The functions margin.table and prop.table also work on multiway tables — the margin argument can be a vector if the relevant margin has two or more dimensions.

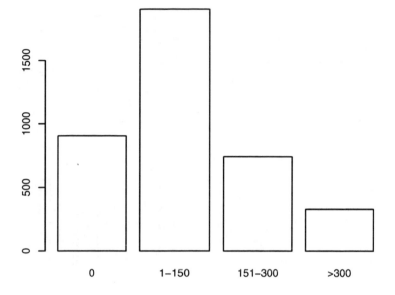

Figure 4.9. Simple `barplot` of total caffeine consumption.

4.6 Graphical display of tables

For presentation purposes, it may be desirable to display a graph rather than a table of counts or percentages. In this section, the main methods for doing this are described.

4.6.1 Barplots

Barplots are made using `barplot`. This function takes an argument, which can be a vector or a matrix. The simplest variant goes as follows (Figure 4.9):

```
> total.caff <- margin.table(caff.marital,2)
> total.caff
consumption
      0   1-150 151-300    >300
    906    1910     742     330
> barplot(total.caff, col="white")
```

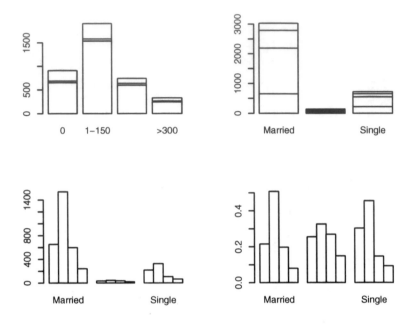

Figure 4.10. Four variants of `barplot` on a two-way table.

Without the `col="white"` argument, the plot comes out in colour, but this is not suitable for a black and white book illustration.

If the argument is a matrix, then `barplot` creates by default a "stacked barplot", where the columns are partitioned according to the contributions from different rows of the table. If you want to place the row contributions beside each other instead, you can use the argument `beside=T`. A series of variants is found in Figure 4.10, which is constructed as follows:

```
> par(mfrow=c(2,2))
> barplot(caff.marital, col="white")
> barplot(t(caff.marital), col="white")
> barplot(t(caff.marital), col="white", beside=T)
> barplot(prop.table(t(caff.marital),2), col="white", beside=T)
> par(mfrow=c(1,1))
```

In the last three plots, we switched rows and columns with the transposition function `t`. In the very last one, the columns are expressed as proportions of the total number in the group. Thus, information is lost on the relative sizes of the marital status groups, but the group of previously married women (recall that the data set deals with women giving birth)

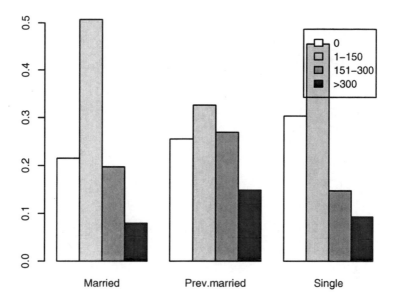

Figure 4.11. Bar plot with specified colours and legend.

is so small that it otherwise becomes almost impossible to compare their caffeine consumption profile with those of the other groups.

As usual, there are a multitude of ways to "prettify" the plots. Here is one possibility (Figure 4.11):

```
> barplot(prop.table(t(caff.marital),2),beside=T,
+ legend.text=colnames(caff.marital),
+ col=c("white","grey80","grey50","black"))
```

Notice that the legend overlaps the top of one of the columns. R is not designed to be able to find a clear area in which to place the legend. However, you can get full control of the legend's position if you insert it explicitly with the `legend` function. For that purpose, it will be helpful to use `locator()`, which allows you to click a mouse button over the plot and have the coordinates returned. See p. 209 for more about this.

4.6.2 Dotcharts

The Cleveland dotcharts, named after William S. Cleveland (1994), can be employed to study a table from both sides at the same time. They contain

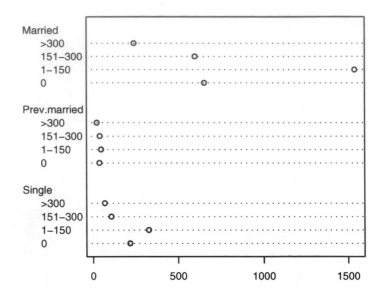

Figure 4.12. Dotchart of caffeine consumption.

the same information as barplots with `beside=T` but give quite a different visual impression. We content ourselves with a single example here (Figure 4.12):

```
> dotchart(t(caff.marital), lcolor="black")
```

(The line colour was changed from the default `"gray"` because it tends to be hard to see in print.)

4.6.3 Piecharts

Piecharts are traditionally frowned upon by statisticians because they are so often used to make trivial data look impressive and are difficult to decode for the human mind. They very rarely contain information that would not have been at least as effectively conveyed in a barplot. Once in a while they are useful, though, and it is no problem to get R to draw them. Here is a way to represent the table of caffeine consumption versus marital status (Figure 4.13; see Section 4.4.3 for an explanation of the "par magic" used to reduce the space between the subplots):

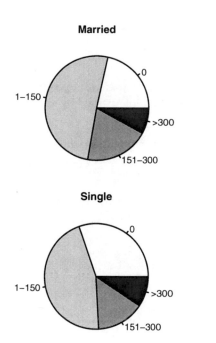

Figure 4.13. Pie charts of caffeine consumption according to marital status.

```
> opar <- par(mfrow=c(2,2),mex=0.8, mar=c(1,1,2,1))
> slices <- c("white","grey80","grey50","black")
> pie(caff.marital["Married",], main="Married", col=slices)
> pie(caff.marital["Prev.married",],
+         main="Previously married", col=slices)
> pie(caff.marital["Single",], main="Single", col=slices)
> par(opar)
```

The `col` argument sets the colour of the pie slices.

There are more possibilities with `piechart`. The help page for `pie` contains an illustrative example concerning the distribution of pie sales (!) by pie type.

4.7 Exercises

4.1 Explore the possibilities for different kinds of line and point plots. Vary the plot symbol, line type, line width, and colour.

4.2   If you make a plot like `plot(rnorm(10),type="o")` with over-plotted lines and points, the lines will be visible inside the plotting symbols. How can this be avoided?

4.3   How can you overlay two `qqnorm` plots in the same plotting area? What goes wrong if you try to generate the plot using `type="l"`, and how do you avoid that?

4.4   Plot a histogram for the `react` data set. Since these data are highly discretized, the histogram will be biased. Why? You may want to try `truehist` from the `MASS` package as a replacement.

4.5   Generate a sample vector `z` of five random numbers from the uniform distribution, and plot `quantile(z,x)` as a function of `x` (use `curve`, for instance).

5

One- and two-sample tests

Most of the rest of this book describes applications of R for actual statistical analysis. The focus to some extent shifts from explanation of the syntax to description of the output and specific arguments to the relevant functions.

Some of the most basic statistical tests deal with comparing continuous data either between two groups or against an a priori stipulated value. This is the topic for this chapter.

Two functions are introduced here, namely t.test and wilcox.test for *t* tests and Wilcoxon tests, respectively. Both can be applied to one- and two-sample problems as well as paired data. Notice that the "two-sample Wilcoxon test" is the same as the one called the "Mann–Whitney test" in many textbooks.

5.1 One-sample *t* test

The *t* tests are based on an assumption that data come from the normal distribution. In the one-sample case we thus have data $x_1, \ldots, x_n$ assumed to be independent realizations of random variables with distribution $N(\mu, \sigma^2)$, which denotes the normal distribution with mean μ and variance σ^2, and we wish to test the *null hypothesis* that $\mu = \mu_0$. We can estimate the parameters μ and σ by the empirical mean $\bar{x}$ and standard

P. Dalgaard, *Introductory Statistics with R*,
DOI: 10.1007/978-0-387-79054-1_5, © Springer Science+Business Media, LLC 2008

deviation s, although we must realize that we could never pinpoint their values exactly.

The key concept is that of the *standard error of the mean*, or SEM. This describes the variation of the average of n random values with mean μ and variance σ^2. This value is

$$\text{SEM} = \sigma/\sqrt{n}$$

and means that if you were to repeat the entire experiment several times and calculate an average for each experiment, then these averages would follow a distribution that is narrower than that of the original distribution. The crucial point is that even based on a single sample, it is possible to calculate an empirical SEM as $s/\sqrt{n}$ using the empirical standard deviation of the sample. This value will tell us how far the observed mean may reasonably have strayed from its true value. For normally distributed data, the rule of thumb is that there is a 95% probability of staying within $\mu \pm 2\sigma$, so we would expect that if μ_0 were the true mean, then $\bar{x}$ should be within 2 SEMs of it. Formally, you calculate

$$t = \frac{\bar{x} - \mu_0}{\text{SEM}}$$

and see whether this falls within an *acceptance region* outside which t should fall with probability equal to a specified *significance level*. This is often chosen as 5%, in which case the acceptance region is almost, but not exactly, the interval from -2 to 2.

In small samples, it is necessary to correct for the fact that an empirical SEM is used and that the distribution of t therefore has somewhat "heavier tails" than the $N(0,1)$: Large deviations happen more frequently than in the normal distribution since they can result from normalizing with an SEM that is too small. The correct values for the acceptance region can be looked up as quantiles in the t distribution with $f = n - 1$ degrees of freedom.

If t falls outside the acceptance region, then we reject the null hypothesis at the chosen significance level. Alternatively (and equivalently), you can calculate the *p-value*, which is the probability of obtaining a value as numerically large as or larger than the observed t and reject the hypothesis if the p-value is less than the significance level.

Sometimes you have prior information on the direction of an effect; for instance, that all plausible mechanisms that would cause μ not to equal μ_0 would tend to make it bigger. In those cases, you may choose to reject the hypothesis only if t falls in the upper tail of the distribution. This is known as *testing against a one-sided alternative*. Since removing the lower tail from the rejection region effectively halves the significance level, a one-sided test at a given level will have a smaller cutoff point. Similarly, p-values are calculated as the probability of a larger value than observed rather than a numerically larger one, effectively halving the *p*-value as long as the observed effect is in the stipulated direction. One-sided tests should be used with some care, preferably only when there is a clear statement of the intent to use them in the study protocol. Switching to a one-sided test to make an otherwise nonsignificant result significant could easily be regarded as dishonest.

Here is an example concerning daily energy intake in kJ for 11 women (Altman, 1991, p. 183). First, the values are placed in a data vector:

```
> daily.intake <- c(5260,5470,5640,6180,6390,6515,
+ 6805,7515,7515,8230,8770)
```

Let us first look at some simple summary statistics, even though these are hardly necessary for such a small data set:

```
> mean(daily.intake)
[1] 6753.636
> sd(daily.intake)
[1] 1142.123
> quantile(daily.intake)
   0%   25%   50%   75%  100%
5260 5910 6515 7515 8770
```

You might wish to investigate whether the women's energy intake deviates systematically from a recommended value of 7725 kJ. Assuming that data come from a normal distribution, the object is to test whether this distribution might have mean $\mu = 7725$. This is done with t.test as follows:

```
> t.test(daily.intake,mu=7725)

        One Sample t-test

data:  daily.intake
t = -2.8208, df = 10, p-value = 0.01814
alternative hypothesis: true mean is not equal to 7725
95 percent confidence interval:
 5986.348 7520.925
sample estimates:
mean of x
 6753.636
```

This is an example of the exact same type as used in the introductory Section 1.1.4. The description of the output is quite superficial there. Here it is explained more thoroughly.

The layout is common to many of the standard statistical tests, and a "dissection" is given in the following:

```
One Sample t-test
```

This should be self-explanatory. It is simply a description of the test that we have asked for. Notice that, by looking at the format of the function call, t.test has automatically found out that a one-sample test is desired.

```
data:   daily.intake
```

This tells us which data are being tested. Of course, this will be obvious *unless* output has been separated from the command that generated it. This can happen, for example, when using the source function to read commands from an external file.

```
t = -2.8208, df = 10, p-value = 0.01814
```

This is where it begins to get interesting. We get the t statistic, the associated degrees of freedom, and the exact p-value. We do not need to use a table of the t distribution to look up which quantiles the t-value can be found between. You can immediately see that $p < 0.05$ and thus that (using the customary 5% level of significance) data deviate significantly from the hypothesis that the mean is 7725.

```
alternative hypothesis: true mean is not equal to 7725
```

This contains two important pieces of information: (a) the value we wanted to test whether the mean could be equal to (7725 kJ) and (b) that the test is two-sided ("not equal to").

```
95 percent confidence interval:
 5986.348 7520.925
```

This is a 95% confidence interval for the true mean; that is, the set of (hypothetical) mean values from which the data do not deviate significantly. It is based on inverting the t test by solving for the values of μ_0 that cause t to lie within its acceptance region. For a 95% confidence interval, the solution is

$$\bar{x} - t_{0.975}(f) \times \text{SEM} < \mu < \bar{x} + t_{0.975}(f) \times \text{SEM}$$

```
sample estimates:
mean of x
 6753.636
```

This final item is the observed mean; that is, the (point) estimate of the true mean.

The function t.test has a number of optional arguments, three of which are relevant in one-sample problems. We have already seen the use of mu to specify the mean value μ under the null hypothesis (default is mu=0). In addition, you can specify that a one-sided test is desired against alternatives greater than μ by using alternative="greater" or alternatives less than μ using alternative="less". The third item that can be specified is the *confidence level* used for the confidence intervals; you would write conf.level=0.99 to get a 99% interval.

Actually, it is often allowable to abbreviate a longish argument specification; for instance, it is sufficient to write alt="g" to get the test against greater alternatives.

5.2   Wilcoxon signed-rank test

The t tests are fairly robust against departures from the normal distribution especially in larger samples, but sometimes you wish to avoid making that assumption. To this end, the *distribution-free methods* are convenient. These are generally obtained by replacing data with corresponding order statistics.

For the one-sample Wilcoxon test, the procedure is to subtract the theoretical μ_0 and rank the differences according to their numerical value, ignoring the sign, and then calculate the sum of the positive or negative ranks. The point is that, assuming only that the distribution is symmetric around μ_0, the test statistic corresponds to selecting each number from 1 to n with probability $1/2$ and calculating the sum. The distribution of the test statistic can be calculated exactly, at least in principle. It becomes computationally excessive in large samples, but the distribution is then very well approximated by a normal distribution.

Practical application of the Wilcoxon signed-rank test is done almost exactly like the t test:

```
> wilcox.test(daily.intake, mu=7725)
        Wilcoxon signed rank test with continuity correction

data:  daily.intake
V = 8, p-value = 0.0293
alternative hypothesis: true location is not equal to 7725

Warning message:
In wilcox.test.default(daily.intake, mu = 7725) :
  cannot compute exact p-value with ties
```

There is not quite as much output as from t.test due to the fact that there is no such thing as a parameter estimate in a nonparametric test and therefore no confidence limits, etc., either. It is, however, possible under some assumptions to define a location measure and calculate confidence intervals for it. See the help files for `wilcox.test` for details.

The relative merits of distribution-free (or *nonparametric*) versus parametric methods such as the *t* test are a contentious issue. If the model assumptions of the parametric test are fulfilled, then it will be somewhat more efficient, on the order of 5% in large samples, although the difference can be larger in small samples. Notice, for instance, that unless the sample size is 6 or above, the signed-rank test simply cannot become significant at the 5% level. This is probably not too important, though; what is more important is that the apparent lack of assumptions for these tests sometimes misleads people into using them for data where the observations are not independent or where a comparison is biased by an important covariate.

The Wilcoxon tests are susceptible to the problem of *ties*, where several observations share the same value. In such cases, you simply use the average of the tied ranks; for example, if there are four identical values corresponding to places 6 to 9, they will all be assigned the value 7.5. This is not a problem for the large-sample normal approximations, but the exact small-sample distributions become much more difficult to calculate and `wilcox.test` cannot do so.

The test statistic V is the sum of the positive ranks. In the example, the *p*-value is computed from the normal approximation because of the tie at 7515.

The function `wilcox.test` takes arguments `mu` and `alternative`, just like `t.test`. In addition, it has `correct`, which turns a continuity correction on or off (the default is "on", as seen from the output title; `correct=F` turns it off), and `exact`, which specifies whether exact tests should be calculated. Recall that "on/off" options such as these are specified using logical values that can be either `TRUE` or `FALSE`.

5.3   Two-sample *t* test

The two-sample *t* test is used to test the hypothesis that two samples may be assumed to come from distributions with the same mean.

The theory for the two-sample *t* test is not very different in principle from that of the one-sample test. Data are now from two groups, $x_{11}, \ldots, x_{1n_1}$ and $x_{21}, \ldots, x_{2n_2}$, which we assume are sampled from the normal distributions $N(\mu_1, \sigma_1^2)$ and $N(\mu_2, \sigma_2^2)$, and it is desired to test the null hypothesis $\mu_1 = \mu_2$. You then calculate

$$t = \frac{\bar{x}_2 - \bar{x}_1}{\text{SEDM}}$$

where the *standard error of difference of means* is

$$\text{SEDM} = \sqrt{\text{SEM}_1^2 + \text{SEM}_2^2}$$

There are two ways of calculating the SEDM depending on whether or not you assume that the two groups have the same variance. The "classical" approach is to assume that the variances are identical. With this approach, you first calculate a pooled *s* based on the standard deviations from the two groups and plug that value into the SEM. Under the null hypothesis, the *t* value will follow a *t* distribution with $n_1 + n_2 - 2$ degrees of freedom.

An alternative procedure due to Welch is to calculate the SEMs from the separate group standard deviations s_1 and s_2. With this procedure, *t* is actually not *t*-distributed, but its distribution may be approximated by a *t* distribution with a number of degrees of freedom that can be calculated from s_1, s_2, and the group sizes. This is generally not an integer.

The Welch procedure is generally considered the safer one. Usually, the two procedures give very similar results unless both the group sizes and the standard deviations are very different.

We return to the daily energy expenditure data (see Section 1.2.14) and consider the problem of comparing energy expenditures between lean and obese women.

```
> attach(energy)
> energy
    expend stature
1     9.21   obese
2     7.53    lean
3     7.48    lean
...
20    7.58    lean
21    9.19   obese
22    8.11    lean
```

Notice that the necessary information is contained in two parallel columns of a data frame. The factor `stature` contains the group and the numeric variable `expend` the energy expenditure in mega-Joules. R allows data in this format to be analyzed by `t.test` and `wilcox.test` using a model formula specification. An older format (still available) requires you to specify data from each group in a separate variable, but the newer format is much more convenient for data that are kept in data frames and is also more flexible if you later want to group the same response data according to other criteria.

The object is to see whether there is a shift in level between the two groups, so we apply a *t* test as follows:

```
> t.test(expend~stature)

        Welch Two Sample t-test

data:   expend by stature
t = -3.8555, df = 15.919, p-value = 0.001411
alternative hypothesis: true difference in means is not equal to 0
95 percent confidence interval:
 -3.459167 -1.004081
sample estimates:
 mean in group lean mean in group obese
         8.066154            10.297778
```

Notice the use of the tilde (~) operator to specify that expend is *described by* stature.

The output is not much different from that of the one-sample test. The confidence interval is for the *difference* in means and does not contain 0, which is in accordance with the *p*-value indicating a significant difference at the 5% level.

It is Welch's variant of the *t* test that is calculated by default. This is the test where you do not assume that the variance is the same in the two groups, which (among other things) results in the fractional degrees of freedom.

To get the usual (textbook) *t* test, you must specify that you are willing to assume that the variances are the same. This is done via the optional argument var.equal=T; that is:

```
> t.test(expend~stature, var.equal=T)

        Two Sample t-test

data:   expend by stature
t = -3.9456, df = 20, p-value = 0.000799
alternative hypothesis: true difference in means is not equal to 0
95 percent confidence interval:
 -3.411451 -1.051796
sample estimates:
 mean in group lean mean in group obese
         8.066154            10.297778
```

Notice that the degrees of freedom now has become a whole number, namely $13 + 9 - 2 = 20$. The *p*-value has dropped slightly (from 0.14% to 0.08%) and the confidence interval is a little narrower, but overall the changes are slight.

5.4   Comparison of variances

Even though it is possible in R to perform the two-sample t test without the assumption that the variances are the same, you may still be interested in testing that assumption, and R provides the var.test function for that purpose, implementing an F test on the ratio of the group variances. It is called the same way as t.test:

```
> var.test(expend~stature)

        F test to compare two variances

data:  expend by stature
F = 0.7844, num df = 12, denom df =  8, p-value = 0.6797
alternative hypothesis: true ratio of variances is not equal to 1
95 percent confidence interval:
 0.1867876 2.7547991
sample estimates:
ratio of variances
          0.784446
```

The test is not significant, so there is no evidence against the assumption that the variances are identical. However, the confidence interval is very wide. For small data sets such as this one, the assumption of constant variance is largely a matter of belief. It may also be noted that this test is not robust against departures from a normal distribution. The stats package contains several alternative tests for variance homogeneity, each with its own assumptions, benefits, and drawbacks, but we shall not discuss them at length.

Notice that the test is based on the assumption that the groups are independent. You should not apply this test to paired data.

5.5   Two-sample Wilcoxon test

You might prefer a nonparametric test if you doubt the normal distribution assumptions of the t test. The two-sample Wilcoxon test is based on replacing the data by their rank (without regard to grouping) and calculating the sum of the ranks in one group, thus reducing the problem to one of sampling n_1 values without replacement from the numbers 1 to $n_1 + n_2$.

This is done using wilcox.test, which behaves similarly to t.test:

```
> wilcox.test(expend~stature)

        Wilcoxon rank sum test with continuity correction

data:   expend by stature
W = 12, p-value = 0.002122
alternative hypothesis: true location shift is not equal to 0

Warning message:
In wilcox.test.default(x = c(7.53, 7.48, 8.08, 8.09, 10.15, 8.4,   :
  cannot compute exact p-value with ties
```

The test statistic W is the sum of ranks in the first group minus its theoretical minimum (i.e., it is zero if all the smallest values fall in the first group). Some textbooks use a statistic that is the sum of ranks in the *smallest* group with no minimum correction, which is of course equivalent. Notice that, as in the one-sample example, we are having problems with ties and rely on the approximate normal distribution of W.

5.6   The paired t test

Paired tests are used when there are two measurements on the same experimental unit. The theory is essentially based on taking differences and thus reducing the problem to that of a one-sample test. Notice, though, that it is implicitly assumed that such differences have a distribution that is independent of the level. A useful graphical check is to make a scatterplot of the pairs with the line of identity added or to plot the difference against the average of the pair (sometimes called a *Bland–Altman plot*). If there seems to be a tendency for the dispersion to change with the level, then it may be useful to transform the data; frequently the standard deviation is proportional to the level, in which case a logarithmic transformation is useful.

The data on pre- and postmenstrual energy intake in a group of women are considered several times in Chapter 1 (and you may notice that the first column is identical to `daily.intake`, which was used in Section 5.1). There data are entered from the command line, but they are also available as a data set in the ISwR package:

```
> attach(intake)
> intake
    pre post
1  5260 3910
2  5470 4220
3  5640 3885
4  6180 5160
```

```
5   6390 5645
6   6515 4680
7   6805 5265
8   7515 5975
9   7515 6790
10  8230 6900
11  8770 7335
```

The point is that the same 11 women are measured twice, so it makes sense to look at individual differences:

```
> post - pre
 [1] -1350 -1250 -1755 -1020  -745 -1835 -1540 -1540  -725 -1330
[11] -1435
```

It is immediately seen that they are all negative. All the women have a lower energy intake postmenstrually than premenstrually. The paired *t* test is obtained as follows:

```
> t.test(pre, post, paired=T)

        Paired t-test

data:  pre and post
t = 11.9414, df = 10, p-value = 3.059e-07
alternative hypothesis: true difference in means is not equal to 0
95 percent confidence interval:
 1074.072 1566.838
sample estimates:
mean of the differences
               1320.455
```

There is not much new to say about the output; it is virtually identical to that of a one-sample *t* test on the elementwise differences.

Notice that you have to specify `paired=T` explicitly in the call, indicating that you want a paired test. In the old-style interface for the unpaired *t* test, the two groups are specified as separate vectors and you would request that analysis by omitting `paired=T`. If data are actually paired, then it would be seriously inappropriate to analyze them without taking the pairing into account.

Even though it might be considered pedagogically dubious to show what you should *not* do, the following shows the results of an unpaired *t* test on the same data for comparison:

```
> t.test(pre, post) #WRONG!

          Welch Two Sample t-test

data:  pre and post
t = 2.6242, df = 19.92, p-value = 0.01629
alternative hypothesis: true difference in means is not equal to 0
95 percent confidence interval:
  270.5633 2370.3458
sample estimates:
mean of x mean of y
 6753.636   5433.182
```

The number symbol (or "hash") # introduces a comment in R. The rest of the line is skipped.

It is seen that t has become considerably smaller, although still significant at the 5% level. The confidence interval has become almost four times wider than in the correct paired analysis. Both illustrate the loss of efficiency caused by not using the information that the "pre" and "post" measurements are from the same person. Alternatively, you could say that it demonstrates the gain in efficiency obtained by planning the experiment with two measurements on the same person, rather than having two independent groups of pre- and postmenstrual women.

5.7   The matched-pairs Wilcoxon test

The paired Wilcoxon test is the same as a one-sample Wilcoxon signed-rank test on the differences. The call is completely analogous to t.test:

```
> wilcox.test(pre, post, paired=T)
           Wilcoxon signed rank test with continuity correction

data:  pre and post
V = 66, p-value = 0.00384
alternative hypothesis: true location shift is not equal to 0

Warning message:
In wilcox.test.default(pre, post, paired = T) :
  cannot compute exact p-value with ties
```

The result does not show any material difference from that of the t test. The p-value is not quite so extreme, which is not too surprising since the Wilcoxon rank sum cannot get any larger than it does when all differences have the same sign, whereas the t statistic can become arbitrarily extreme.

Again, we have trouble with tied data invalidating the exact p calculations. This time it is the two identical differences of -1540.

In the present case it is actually very easy to calculate the exact p-value for the Wilcoxon test. It is the probability of 11 positive differences + the probability of 11 negative ones, $2 \times (1/2)^{11} = 1/1024 = 0.00098$, so the approximate p-value is almost four times too large.

5.8   Exercises

5.1   Do the values of the `react` data set (notice that this is a single vector, not a data frame) look reasonably normally distributed? Does the mean differ significantly from zero according to a t test?

5.2   In the data set `vitcap`, use a t test to compare the vital capacity for the two groups. Calculate a 99% confidence interval for the difference. The result of this comparison may be misleading. Why?

5.3   Perform the analyses of the `react` and `vitcap` data using nonparametric techniques.

5.4   Perform graphical checks of the assumptions for a paired t test in the `intake` data set.

5.5   The function `shapiro.test` computes a test of normality based on the degree of linearity of the Q–Q plot. Apply it to the `react` data. Does it help to remove the outliers?

5.6   The crossover trial in `ashina` can be analyzed for a drug effect in a simple way (how?) if you ignore a potential period effect. However, you can do better. Hint: Consider the intra-individual differences; if there were *only* a period effect present, how should the differences behave in the two groups? Compare the results of the simple method and the improved method.

5.7   Perform 10 one-sample t tests on simulated normally distributed data sets of 25 observations each. Repeat the experiment, but instead simulate samples from a different distribution; try the t distribution with 2 degrees of freedom and the exponential distribution (in the latter case, test for the mean being equal to 1). Can you find a way to automate this so that you can have a larger number of replications?

6

Regression and correlation

The main object of this chapter is to show how to perform basic regression analyses, including plots for model checking and display of confidence and prediction intervals. Furthermore, we describe the related topic of correlation in both its parametric and nonparametric variants.

6.1 Simple linear regression

We consider situations where you want to describe the relation between two variables using linear regression analysis. You may, for instance, be interested in describing short.velocity as a function of blood.glucose. This section deals only with the very basics, whereas several more complicated issues are postponed until Chapter 12.

The linear regression model is given by

$$y_i = \alpha + \beta x_i + \epsilon_i$$

in which the ϵ_i are assumed independent and $N(0, \sigma^2)$. The nonrandom part of the equation describes the y_i as lying on a straight line. The slope of the line (the *regression coefficient*) is β, the increase per unit change in x. The line intersects the y-axis at the *intercept* α.

The parameters α, β, and σ^2 can be estimated using the *method of least squares*. Find the values of α and β that minimize the sum of squared

P. Dalgaard, *Introductory Statistics with R*,
DOI: 10.1007/978-0-387-79054-1_6, © Springer Science+Business Media, LLC 2008

residuals

$$SS_{res} = \sum_i (y_i - (\alpha + \beta x_i))^2$$

This is not actually done by trial and error. One can find closed-form expressions for the choice of parameters that gives the smallest value of SS_{res}:

$$\hat{\beta} = \frac{\sum(x_i - \bar{x})(y_i - \bar{y})}{\sum(x_i - \bar{x})^2}$$

$$\hat{\alpha} = \bar{y} - \hat{\beta}\bar{x}$$

The residual variance is estimated as $SS_{res}/(n-2)$, and the residual standard deviation is of course the square root of that.

The empirical slope and intercept will deviate somewhat from the true values due to sampling variation. If you were to generate several sets of y_i at the same set of x_i, you would observe a distribution of empirical slopes and intercepts. Just as you could calculate the SEM to describe the variability of the empirical mean, it is also possible from a single sample of (x_i, y_i) to calculate the standard error of the computed estimates, s.e.$(\hat{\alpha})$ and s.e.$(\hat{\beta})$. These standard errors can be used to compute confidence intervals for the parameters and tests for whether a parameter has a specific value.

It is usually of prime interest to test the null hypothesis that $\beta = 0$ since that would imply that the line was horizontal and thus that the ys have a distribution that is the same, whatever the value of x. You can compute a t test for that hypothesis simply by dividing the estimate by its standard error

$$t = \frac{\hat{\beta}}{\text{s.e.}(\hat{\beta})}$$

which follows a t distribution on $n-2$ degrees of freedom if the true β is zero. A similar test can be calculated for whether the intercept is zero, but you should be aware that it is often a meaningless hypothesis either because there is no natural reason to believe that the line should go through the origin or because it would involve an extrapolation far outside the range of data.

For the example in this section, we need the data frame thuesen, which we attach with

```
> attach(thuesen)
```

For linear regression analysis, the function lm (*linear model*) is used:

```
> lm(short.velocity~blood.glucose)

Call:
lm(formula = short.velocity ~ blood.glucose)

Coefficients:
  (Intercept)   blood.glucose
      1.09781         0.02196
```

The argument to lm is a *model formula* in which the tilde symbol (~) should be read as "described by". This was seen several times earlier, both in connection with boxplots and stripcharts and with the *t* and Wilcoxon tests.

The lm function handles much more complicated models than simple linear regression. There can be many other things besides a dependent and a descriptive variable in a model formula. A multiple linear regression analysis (which we discuss in Chapter 11) of, for example, y on x1, x2, and x3 is specified as y ~ x1 + x2 + x3.

In its raw form, the output of lm is very brief. All you see is the estimated intercept (α) and the estimated slope (β). The best-fitting straight line is seen to be short.velocity $= 1.098 + 0.0220 \times$ blood.glucose, but for instance no tests of significance are given.

The result of lm is a *model object*. This is a distinctive concept of the S language (of which R is a dialect). Whereas other statistical systems focus on generating printed output that can be controlled by setting options, you get instead the result of a model fit encapsulated in an object from which the desired quantities can be obtained using *extractor functions*. An lm object does in fact contain much more information than you see when it is printed.

A basic extractor function is summary:

```
> summary(lm(short.velocity~blood.glucose))

Call:
lm(formula = short.velocity ~ blood.glucose)

Residuals:
     Min       1Q    Median       3Q       Max
-0.40141  -0.14760  -0.02202  0.03001  0.43490

Coefficients:
               Estimate Std. Error t value Pr(>|t|)
(Intercept)     1.09781    0.11748   9.345 6.26e-09 ***
blood.glucose   0.02196    0.01045   2.101   0.0479 *
---
Signif. codes:  0 '***' 0.001 '**' 0.01 '*' 0.05 '.' 0.1 ' ' 1
```

```
Residual standard error: 0.2167 on 21 degrees of freedom
   (1 observation deleted due to missingness)
Multiple R-squared: 0.1737,      Adjusted R-squared: 0.1343
F-statistic: 4.414 on 1 and 21 DF,  p-value: 0.0479
```

The format above looks more like what other statistical packages would output. The following is a "dissection" of the output:

```
Call:
lm(formula = short.velocity ~ blood.glucose)
```

As in t.test, etc., the output starts with something that is essentially a repeat of the function call. This is not very interesting when one has just given it as a command to R, but it is useful if the result is saved in a variable that is printed later.

```
Residuals:
    Min       1Q    Median        3Q      Max
-0.40141 -0.14760 -0.02202  0.03001  0.43490
```

This gives a superficial view of the distribution of the residuals that may be used as a quick check of the distributional assumptions. The average of the residuals is zero by definition, so the median should not be far from zero, and the minimum and maximum should be roughly equal in absolute value. In the example, it can be noticed that the third quartile is remarkably close to zero, but in view of the small number of observations, this is not really something to worry about.

```
Coefficients:
              Estimate Std. Error t value Pr(>|t|)
(Intercept)    1.09781    0.11748   9.345 6.26e-09 ***
blood.glucose  0.02196    0.01045   2.101   0.0479 *
---
Signif. codes:  0 '***' 0.001 '**' 0.01 '*' 0.05 '.' 0.1 ' ' 1
```

Here we see the regression coefficient and the intercept again, but this time with accompanying standard errors, t tests, and p-values. The symbols to the right are graphical indicators of the level of significance. The line below the table shows the definition of these indicators; one star means $0.01 < p < 0.05$.

The graphical indicators have been the target of some controversy. Some people like to have the possibility of seeing at a glance whether there is "anything interesting" in an analysis, whereas others feel that the indicators too often correspond to meaningless tests. For instance, the intercept in the analysis above is hardly a meaningful quantity at all, and the three-star significance of it is certainly irrelevant. If you are bothered by the stars, turn them off with options(show.signif.stars=FALSE).

```
Residual standard error: 0.2167 on 21 degrees of freedom
  (1 observation deleted due to missingness)
```

This is the residual variation, an expression of the variation of the observations around the regression line, estimating the model parameter σ. The model is not fitted to the entire data set because one value of short.velocity is missing.

```
Multiple R-squared: 0.1737,   Adjusted R-squared: 0.1343
```

The first item above is R^2, which in a simple linear regression may be recognized as the squared Pearson correlation coefficient (see Section 6.4.1); that is, $R^2 = r^2$. The other one is the adjusted R^2; if you multiply it by 100%, it can be interpreted as "% variance reduction" (this can, in fact, become negative).

```
F-statistic: 4.414 on 1 and 21 DF, p-value: 0.0479
```

This is an F test for the hypothesis that the regression coefficient is zero. This test is not really interesting in a simple linear regression analysis since it just duplicates information already given — it becomes more interesting when there is more than one explanatory variable. Notice that it gives the exact same result as the t test for a zero slope. In fact, the F test is identical to the square of the t test: $4.414 = (2.101)^2$. This is true in any model with 1 degree of freedom.

We will see later how to draw residual plots and plots of data with confidence and prediction limits. First, we draw just the points and the fitted line. Figure 6.1 has been constructed as follows:

```
> plot(blood.glucose,short.velocity)
> abline(lm(short.velocity~blood.glucose))
```

abline, meaning (a, b)-line, draws lines based on the intercept and slope, a and b, respectively. It can be used with scalar values as in abline(1.1,0.022), but conveniently it can also extract the information from a linear model fitted to data with lm.

6.2   Residuals and fitted values

We have seen how summary can be used to extract information about the results of a regression analysis. Two further extraction functions are fitted and resid. They are used as follows. For convenience, we first store the value returned by lm under the name lm.velo (short for "velocity", but you could of course use any other name).

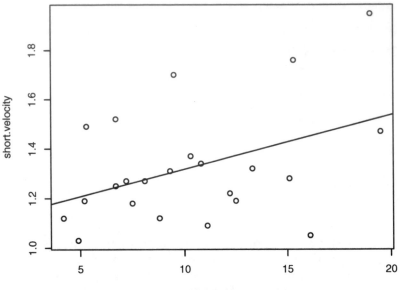

Figure 6.1. Scatterplot with regression line.

```
> lm.velo <- lm(short.velocity~blood.glucose)
> fitted(lm.velo)
        1        2        3        4        5        6        7
1.433841 1.335010 1.275711 1.526084 1.255945 1.214216 1.302066
        8        9       10       11       12       13       14
1.341599 1.262534 1.365758 1.244964 1.212020 1.515103 1.429449
       15       17       18       19       20       21       22
1.244964 1.190057 1.324029 1.372346 1.451411 1.389916 1.205431
       23       24
1.291085 1.306459
> resid(lm.velo)
           1            2            3            4            5
 0.326158532  0.004989882 -0.005711308 -0.056084062  0.014054962
           6            7            8            9           10
 0.275783754  0.007933665 -0.251598875 -0.082533795 -0.145757649
          11           12           13           14           15
 0.005036223 -0.022019994  0.434897199 -0.149448964  0.275036223
          17           18           19           20           21
-0.070057471  0.045971143 -0.182346406 -0.401411486 -0.069916424
          22           23           24
-0.175431237 -0.171085074  0.393541161
```

The function `fitted` returns fitted values — the *y*-values that you would expect for the given *x*-values according to the best-fitting straight line; in the present case, `1.098+0.0220*blood.glucose`. The residuals shown by `resid` is the difference between this and the observed `short.velocity`.

Note that the fitted values and residuals are labelled with the row names of the `thuesen` data frame. Notice in particular that they do not contain observation no. 16, which had a missing value in the response variable.

It is necessary to discuss some awkward aspects that arise when there are missing values in data.

To put the fitted line on the plot, you might, although it is easier to use `abline(lm.velo)`, get the idea of doing it with `lines`, *but*

```
> plot(blood.glucose,short.velocity)
> lines(blood.glucose,fitted(lm.velo))
Error in xy.coords(x, y) : 'x' and 'y' lengths differ
Calls: lines -> lines.default -> plot.xy -> xy.coords
```

which is true. There are 24 observations but only 23 fitted values because one of the `short.velocity` values is NA. Notice, incidentally, that the error occurs within a series of nested function calls, which are being listed along with the error message to reduce confusion.

What we needed was `blood.glucose`, but only for those patients whose `short.velocity` has been recorded.

```
> lines(blood.glucose[!is.na(short.velocity)],fitted(lm.velo))
```

Recall that the `is.na` function yields a vector that is TRUE wherever the argument is NA (missing). One advantage to this method is that the fitted line does not extend beyond the range of data. The technique works but becomes clumsy if there are missing values in several variables:

```
...blood.glucose[!is.na(short.velocity) & !is.na(blood.glucose)]...
```

It becomes easier with the function `complete.cases`, which can find observations that are nonmissing on several variables or across an entire data frame.

```
> cc <- complete.cases(thuesen)
```

We could then attach `thuesen[cc,]` and work on from there. However, there is a better alternative available: You can use the `na.exclude` method for NA handling. This can be set either as an argument to `lm` or as an option; that is,

```
> options(na.action=na.exclude)
> lm.velo <- lm(short.velocity~blood.glucose)
```

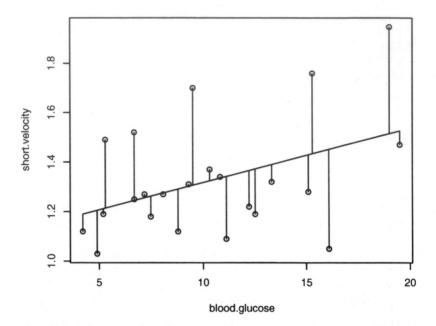

Figure 6.2. Scatterplot of short.velocity versus blood.glucose with fitted line and residual line segments.

```
> fitted(lm.velo)
       1        2        3        4        5        6        7
1.433841 1.335010 1.275711 1.526084 1.255945 1.214216 1.302066
       8        9       10       11       12       13       14
1.341599 1.262534 1.365758 1.244964 1.212020 1.515103 1.429449
      15       16       17       18       19       20       21
1.244964       NA 1.190057 1.324029 1.372346 1.451411 1.389916
      22       23       24
1.205431 1.291085 1.306459
```

Notice how the missing observation, no. 16, now appears in the fitted values with a missing fitted value. It is necessary to recalculate the lm.velo object after changing the option.

To create a plot where residuals are displayed by connecting observations to corresponding points on the fitted line, you can do the following. The final result will look like Figure 6.2. segments draws line segments; its arguments are the endpoint coordinates in the order (x_1, y_1, x_2, y_2).

```
> segments(blood.glucose,fitted(lm.velo),
+          blood.glucose,short.velocity)
```

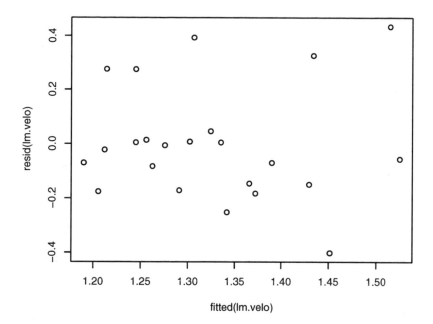

Figure 6.3. `short.velocity` and `blood.glucose`: residuals versus fitted value.

A simple plot of residuals versus fitted values is obtained as (Figure 6.3)

```
> plot(fitted(lm.velo),resid(lm.velo))
```

and we can get an indication of whether residuals might have come from a normal distribution by checking for a straight line on a Q–Q plot (see Section 4.2.3) as follows (Figure 6.4):

```
> qqnorm(resid(lm.velo))
```

6.3   Prediction and confidence bands

Fitted lines are often presented with uncertainty bands around them. There are two kinds of bands, often referred to as the "narrow" and "wide" limits.

The narrow bands, *confidence bands*, reflect the uncertainty about the line itself, like the SEM expresses the precision with which a mean is known. If there are many observations, the bands will be quite narrow, reflecting

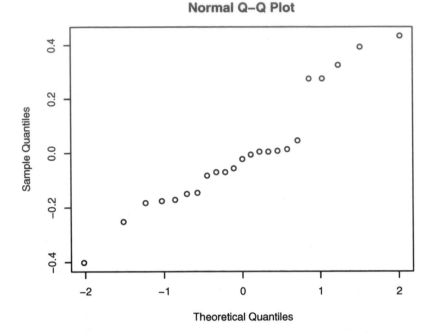

Figure 6.4. `short.velocity` and `blood.glucose`: Q–Q plot of residuals.

a well-determined line. These bands often show a marked curvature since the line is better determined near the center of the point cloud. This is a fact that can be shown mathematically, but you may also understand it intuitively as follows: The predicted value at $\bar{x}$ will be $\bar{y}$, whatever the slope is, and hence the standard error of the fitted value at that point is the SEM of the ys. At other values of x, there will also be a contribution from the variability of the estimated slope, having increasing influence as you move away from $\bar{x}$. Technically, you also need to establish that $\bar{y}$ and $\hat{\beta}$ are uncorrelated.

The wide bands, *prediction bands*, include the uncertainty about future observations. These bands should capture the majority of the observed points and will not collapse to a line as the number of observations increases. Rather, the limits approach the true line ± 2 standard deviations (for 95% limits). In smaller samples, the bands do curve since they include uncertainty about the line itself, but not as markedly as the confidence bands. Obviously, these limits rely strongly on the assumption of normally distributed errors with a constant variance, so you should not use such limits unless you believe that the assumption is a reasonable approximation for the data at hand.

Predicted values, with or without prediction and confidence bands, may be extracted with the function `predict`. With no arguments, it just gives the fitted values:

```
> predict(lm.velo)
       1        2        3        4        5        6        7
1.433841 1.335010 1.275711 1.526084 1.255945 1.214216 1.302066
       8        9       10       11       12       13       14
1.341599 1.262534 1.365758 1.244964 1.212020 1.515103 1.429449
      15       16       17       18       19       20       21
1.244964       NA 1.190057 1.324029 1.372346 1.451411 1.389916
      22       23       24
1.205431 1.291085 1.306459
```

If you add `interval="confidence"` or `interval="prediction"`, then you get the vector of predicted values augmented with limits. The arguments can be abbreviated:

```
> predict(lm.velo,int="c")
        fit      lwr      upr
1  1.433841 1.291371 1.576312
2  1.335010 1.240589 1.429431
...
23 1.291085 1.191084 1.391086
24 1.306459 1.210592 1.402326
> predict(lm.velo,int="p")
        fit       lwr      upr
1  1.433841 0.9612137 1.906469
2  1.335010 0.8745815 1.795439
...
23 1.291085 0.8294798 1.752690
24 1.306459 0.8457315 1.767186
Warning message:
In predict.lm(lm.velo, int = "p") :
  Predictions on current data refer to _future_ responses
```

`fit` denotes the expected values, here identical to the fitted values (they need not be; read on). `lwr` and `upr` are the lower and upper confidence limits for the expected values, respectively, the prediction limits for `short.velocity` for new persons with these values of `blood.glucose`. The warning in this case does not really mean that anything is wrong, but there is a pitfall: The limits should not be used for evaluating the *observed* data to which the line has been fitted. These will tend to lie closer to the line for the extreme x values because those data points are the more influential; that is, the prediction bands curve the wrong way.

The best way to add prediction and confidence intervals to a scatterplot is to use the `matlines` function, which plots the columns of a matrix against a vector.

There are a few snags to this, however: (a) The `blood.glucose` values are in random order; we do not want line segments connecting points haphazardly along the confidence curves; (b) the prediction limits, particularly the lower one, extend outside the plot region; and (c) the `matlines` command needs to be prevented from cycling through line styles and colours. Notice that the `na.exclude` setting (p. 115) prevents us from also having an observation omitted from the predicted values.

The solution is to *predict in a new data frame* containing suitable x values (here `blood.glucose`) at which to predict. It is done as follows:

```
> pred.frame <- data.frame(blood.glucose=4:20)
> pp <- predict(lm.velo, int="p", newdata=pred.frame)
> pc <- predict(lm.velo, int="c", newdata=pred.frame)
> plot(blood.glucose, short.velocity,
+       ylim=range(short.velocity, pp, na.rm=T))
> pred.gluc <- pred.frame$blood.glucose
> matlines(pred.gluc, pc, lty=c(1,2,2), col="black")
> matlines(pred.gluc, pp, lty=c(1,3,3), col="black")
```

What happens is that we create a new data frame in which the variable `blood.glucose` contains the values at which we want predictions to be made. `pp` and `pc` are then made to contain the result of `predict` for the new data in `pred.frame` with prediction limits and confidence limits, respectively.

For the plotting, we first create a standard scatterplot, except that we ensure that it has enough room for the prediction limits. This is obtained by setting `ylim=range(short.velocity, pp, na.rm=T)`. The function `range` returns a vector of length 2 containing the minimum and maximum values of its arguments. We need the `na.rm=T` argument to cause missing values to be skipped for the range computation; notice that `short.velocity` is included to ensure that points outside the prediction limits are not missed (although in this case there are none). Finally, the curves are added, using as x-values the `blood.glucose` used for the prediction and setting the line types and colours to more sensible values. The final result is seen in Figure 6.5.

6.4   Correlation

A correlation coefficient is a symmetric, scale-invariant measure of association between two random variables. It ranges from -1 to $+1$, where the extremes indicate perfect correlation and 0 means no correlation. The sign is negative when large values of one variable are associated with small values of the other and positive if both variables tend to be large or small

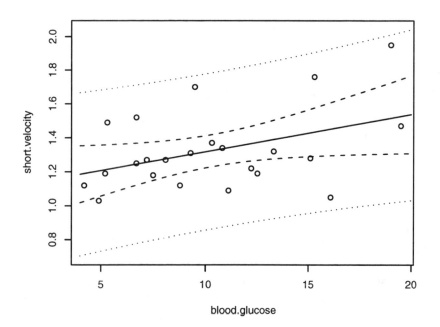

Figure 6.5. Plot with confidence and prediction bands.

simultaneously. The reader should be warned that there are many incorrect uses of correlation coefficients, particularly when they are used in regression-type settings.

This section describes the computation of parametric and nonparametric correlation measures in R.

6.4.1   Pearson correlation

The Pearson correlation is rooted in the two-dimensional normal distribution where the theoretical correlation describes the contour ellipses for the density. If both variables are scaled to have a variance of 1, then a correlation of zero corresponds to circular contours, whereas the ellipses become narrower and finally collapse into a line segment as the correlation approaches ±1.

The empirical correlation coefficient is

$$r = \frac{\sum (x_i - \bar{x})(y_i - \bar{y})}{\sqrt{\sum (x_i - \bar{x})^2 \sum (y_i - \bar{y})^2}}$$

It can be shown that $|r|$ will be less than 1 unless there is a perfect linear relation between x_i and y_i, and for that reason the Pearson correlation is sometimes called the "linear correlation".

It is possible to test the significance of the correlation by transforming it to a t-distributed variable (the formula is not particularly elucidating so we skip it here), which will be identical with the test obtained from testing the significance of the slope of either the regression of y on x or vice versa.

The function cor can be used to compute the correlation between two or more vectors. However, if it is naively applied to the two vectors in thuesen, the following happens:

```
> cor(blood.glucose,short.velocity)
Error in cor(blood.glucose, short.velocity) :
        missing observations in cov/cor
```

All the elementary statistical functions in R require either that all values be nonmissing or that you explicitly state what should be done with the cases with missing values. For mean, var, sd, and similar one-vector functions, you can give the argument na.rm=T to indicate that missing values should be removed before the computation. For cor, you can write

```
> cor(blood.glucose,short.velocity,use="complete.obs")
[1] 0.4167546
```

The reason that cor does not use na.rm=T like the other functions is that there are more possibilities than just removing incomplete cases or failing. If more than two variables are in play, it is also possible to use information from all nonmissing *pairs* of measurements (this might result in a correlation matrix that is not positive definite, though).

You can obtain the entire matrix of correlations between all variables in a data frame by saying, for instance,

```
> cor(thuesen,use="complete.obs")
               blood.glucose short.velocity
blood.glucose      1.0000000      0.4167546
short.velocity     0.4167546      1.0000000
```

Of course, this is more interesting when the data frame contains more than two vectors!

However, the calculations above give no indication of whether the correlation is significantly different from zero. To that end, you need cor.test. It works simply by specifying the two variables:

```
> cor.test(blood.glucose,short.velocity)

        Pearson's product-moment correlation

data:  blood.glucose and short.velocity
t = 2.101, df = 21, p-value = 0.0479
alternative hypothesis: true correlation is not equal to 0
95 percent confidence interval:
 0.005496682 0.707429479
sample estimates:
     cor
0.4167546
```

We also get a confidence interval for the true correlation. Notice that it is exactly the same p-value as in the regression analysis in Section 6.1 and also that based on the ANOVA table for the regression model, which is described in Section 7.5.

6.4.2 Spearman's ρ

As with the one- and two-sample problems, you may be interested in nonparametric variants. These have the advantage of not depending on the normal distribution and, indeed, being invariant to monotone transformations of the coordinates. The main disadvantage is that its interpretation is not quite clear. A popular and simple choice is Spearman's rank correlation coefficient ρ. This is obtained quite simply by replacing the observations by their rank and computing the correlation. Under the null hypothesis of independence between the two variables, the exact distribution of ρ can be calculated.

Unlike group comparisons where there is essentially one function per named test, correlation tests are all grouped into cor.test. There is no special spearman.test function. Instead, the test is considered one of several possibilities for testing correlations and is therefore specified via an option to cor.test:

```
> cor.test(blood.glucose,short.velocity,method="spearman")

        Spearman's rank correlation rho

data:  blood.glucose and short.velocity
S = 1380.364, p-value = 0.1392
alternative hypothesis: true rho is not equal to 0
sample estimates:
     rho
0.318002

Warning message:
```

```
In cor.test.default(blood.glucose, short.velocity, method="spearman"):
  Cannot compute exact p-values with ties
```

6.4.3 Kendall's τ

The third correlation method that you can choose is Kendall's τ, which is based on counting the number of *concordant* and *discordant* pairs. A pair of points is concordant if the difference in the x-coordinate is of the same sign as the difference in the y-coordinate. For a perfect monotone relation, either all pairs will be concordant or all pairs will be discordant. Under independence, there should be as many concordant pairs as there are discordant ones.

Since there are many pairs of points to check, this is quite a computationally intensive procedure compared with the two others. In small data sets such as the present one, it does not matter at all, though, and the procedure is generally usable up to at least 5000 observations.

The τ coefficient has the advantage of a more direct interpretation over Spearman's ρ, but apart from that there is little reason to prefer one over the other.

```
> cor.test(blood.glucose,short.velocity,method="kendall")

        Kendall's rank correlation tau

data:  blood.glucose and short.velocity
z = 1.5604, p-value = 0.1187
alternative hypothesis: true tau is not equal to 0
sample estimates:
      tau
0.2350616

Warning message:
In cor.test.default(blood.glucose, short.velocity, method="kendall"):
  Cannot compute exact p-value with ties
```

Notice that neither of the two nonparametric correlations is significant at the 5% level, which the Pearson correlation is, albeit only borderline significant.

6.5 Exercises

6.1 With the rmr data set, plot metabolic rate versus body weight. Fit a linear regression model to the relation. According to the fitted model, what is the predicted metabolic rate for a body weight of 70 kg? Give a 95% confidence interval for the slope of the line.

6.2   In the `juul` data set, fit a linear regression model for the square root of the IGF-I concentration versus age to the group of subjects over 25 years old.

6.3   In the `malaria` data set, analyze the log-transformed antibody level versus age. Make a plot of the relation. Do you notice anything peculiar?

6.4   One can generate simulated data from the two-dimensional normal distribution with a correlation of ρ by the following technique: (a) Generate X as a normal variate with mean 0 and standard deviation 1; (b) generate Y with mean ρX and standard deviation $\sqrt{1-\rho^2}$. Use this to create scatterplots of simulated data with a given correlation. Compute the Spearman and Kendall statistics for some of these data sets.

7

Analysis of variance and the Kruskal–Wallis test

In this section, we consider comparisons among more than two groups parametrically, using analysis of variance, as well as nonparametrically, using the Kruskal–Wallis test. Furthermore, we look at two-way analysis of variance in the case of one observation per cell.

7.1 One-way analysis of variance

We start this section with a brief sketch of the theory underlying the one-way analysis of variance. A little bit of notation is necessary. Let x_{ij} denote observation no. j in group i, so that x_{35} is the fifth observation in group 3; $\bar{x}_i$ is the mean for group i, and $\bar{x}.$ is the grand mean (average of all observations).

We can decompose the observations as

$$x_{ij} = \bar{x}. + \underbrace{(\bar{x}_i - \bar{x}.)}_{\substack{\text{deviation of} \\ \text{group mean from} \\ \text{grand mean}}} + \underbrace{(x_{ij} - \bar{x}_i)}_{\substack{\text{deviation of} \\ \text{observation from} \\ \text{group mean}}}$$

informally corresponding to the model

$$X_{ij} = \mu + \alpha_i + \epsilon_{ij}, \qquad \epsilon_{ij} \sim N(0, \sigma^2)$$

P. Dalgaard, *Introductory Statistics with R*,
DOI: 10.1007/978-0-387-79054-1_7, © Springer Science+Business Media, LLC 2008

in which the hypothesis that all the groups are the same implies that all α_i are zero. Notice that the error terms ϵ_{ij} are assumed to be independent and have the same variance.

Now consider the sums of squares of the underbraced terms, known as *variation within groups*

$$\text{SSD}_W = \sum_i \sum_j (x_{ij} - \bar{x}_i)^2$$

and *variation between groups*

$$\text{SSD}_B = \sum_i \sum_j (\bar{x}_i - \bar{x}.)^2 = \sum_i n_i (\bar{x}_i - \bar{x}.)^2$$

It is possible to prove that

$$\text{SSD}_B + \text{SSD}_W = \text{SSD}_{\text{total}} = \sum_i \sum_j (x_{ij} - \bar{x}.)^2$$

That is, the total variation is split into a term describing differences between group means and a term describing differences between individual measurements within the groups. One says that the grouping explains part of the total variation, and obviously an informative grouping will explain a large part of the variation.

However, the sums of squares can only be positive, so even a completely irrelevant grouping will always "explain" some part of the variation. The question is how small an amount of explained variation can be before it might as well be due to chance. It turns out that in the absence of any systematic differences between the groups, you should expect the sum of squares to be partitioned according to the degrees of freedom for each term, $k - 1$ for SSD_B and $N - k$ for SSD_W, where k is the number of groups and N is the total number of observations.

Accordingly, you can normalize the sums of squares by calculating *mean squares*:

$$\text{MS}_W = \text{SSD}_W / (N - k)$$
$$\text{MS}_B = \text{SSD}_B / (k - 1)$$

MS_W is the pooled variance obtained by combining the individual group variances and thus an estimate of σ^2. In the absence of a true group effect, MS_B will also be an estimate of σ^2, but if there *is* a group effect, then the differences between group means and hence MS_B will tend to be larger. Thus, a test for significant differences between the group means can be performed by comparing two variance estimates. This is why the procedure is called *analysis of variance* even though the objective is to compare the group means.

A formal test needs to account for the fact that random variation will cause some difference in the mean squares. You calculate

$$F = MS_B/MS_W$$

so that F is ideally 1, but some variation around that value is expected. The distribution of F under the null hypothesis is an F distribution with $k - 1$ and $N - k$ degrees of freedom. You reject the hypothesis of identical means if F is larger than the 95% quantile in that F distribution (if the significance level is 5%). Notice that this test is one-sided; a very small F would occur if the group means were very similar, and that will of course not signify a difference between the groups.

Simple analyses of variance can be performed in R using the function lm, which is also used for regression analysis. For more elaborate analyses, there are also the functions aov and lme (linear mixed effects models, from the nlme package). An implementation of Welch's procedure, relaxing the assumption of equal variances and generalizing the unequal-variance t test, is implemented in oneway.test (see Section 7.1.2).

The main example in this section is the "red cell folate" data from Altman (1991, p. 208). To use lm, it is necessary to have the data values in one vector and a factor variable (see Section 1.2.8) describing the division into groups. The red.cell.folate data set contains a data frame in the proper format.

```
> attach(red.cell.folate)
> summary(red.cell.folate)
      folate              ventilation
 Min.    :206.0    N2O+O2,24h:8
 1st Qu.:249.5     N2O+O2,op :9
 Median :274.0     O2,24h    :5
 Mean    :283.2
 3rd Qu.:305.5
 Max.    :392.0
```

Recall that summary applied to a data frame gives a short summary of the distribution of each of the variables contained in it. The format of the summary is different for numeric vectors and factors, so that provides a check that the variables are defined correctly.

The category names for ventilation mean "N_2O and O_2 for 24 hours", "N_2O and O_2 during operation", and "only O_2 for 24 hours".

In the following, the analysis of variance is demonstrated first and then a couple of useful techniques for the presentation of grouped data as tables and graphs are shown.

The specification of a one-way analysis of variance is analogous to a regression analysis. The only difference is that the descriptive variable needs to be a factor and not a numeric variable. We calculate a model object using `lm` and extract the analysis of variance table with `anova`.

```
> anova(lm(folate~ventilation))
Analysis of Variance Table

Response: folate
            Df Sum Sq Mean Sq F value  Pr(>F)
ventilation  2  15516    7758  3.7113 0.04359 *
Residuals   19  39716    2090
---
Signif. codes:  0 `***' 0.001 `**' 0.01 `*' 0.05 `.' 0.1 ` ' 1
```

Here we have SSD_B and MS_B in the top line and SSD_W and MS_W in the second line.

In statistics textbooks, the sums of squares are most often labelled "between groups" and "within groups". Like most other statistical software, R uses slightly different labelling. Variation between groups is labelled by the name of the grouping factor (`ventilation`), and variation within groups is labelled `Residual`. ANOVA tables can be used for a wide range of statistical models, and it is convenient to use a format that is less linked to the particular problem of comparing groups.

For a further example, consider the data set `juul`, introduced in Section 4.1. Notice that the `tanner` variable in this data set is a numeric vector and not a factor. For purposes of tabulation, this makes little difference, but it would be a serious error to use it in this form in an analysis of variance:

```
> attach(juul)
> anova(lm(igf1~tanner))   ## WRONG!
Analysis of Variance Table

Response: igf1
            Df    Sum Sq  Mean Sq F value    Pr(>F)
tanner       1  10985605 10985605  686.07 < 2.2e-16 ***
Residuals  790  12649728    16012
---
Signif. codes:  0 `***' 0.001 `**' 0.01 `*' 0.05 `.' 0.1 ` ' 1
```

This does not describe a grouping of data but a linear regression on the group number! Notice the telltale 1 DF for the effect of `tanner`.

Things can be fixed as follows:

```
> juul$tanner <- factor(juul$tanner,
+                       labels=c("I","II","III","IV","V"))
```

```
> detach(juul)
> attach(juul)
> summary(tanner)
   I   II  III   IV    V NA's
 515  103   72   81  328  240
> anova(lm(igf1~tanner))
Analysis of Variance Table

Response: igf1
           Df   Sum Sq  Mean Sq F value    Pr(>F)
tanner      4 12696217  3174054  228.35 < 2.2e-16 ***
Residuals 787 10939116    13900
---
Signif. codes:  0 '***' 0.001 '**' 0.01 '*' 0.05 '.' 0.1 ' ' 1
```

We needed to reattach the `juul` data frame in order to use the changed definition. An attached data frame is effectively a separate copy of it (although it does not take up extra space as long as the original is unchanged). The `Df` column now has an entry of 4 for `tanner`, as it should.

7.1.1   Pairwise comparisons and multiple testing

If the *F* test shows that there is a difference between groups, the question quickly arises of where the difference lies. It becomes necessary to compare the individual groups.

Part of this information can be found in the regression coefficients. You can use `summary` to extract regression coefficients with standard errors and *t* tests. These coefficients do not have their usual meaning as the slope of a regression line but have a special interpretation, which is described below.

```
> summary(lm(folate~ventilation))
Call:
lm(formula = folate ~ ventilation)
Residuals:
    Min      1Q  Median      3Q     Max
-73.625 -35.361  -4.444  35.625  75.375
Coefficients:
                   Estimate Std. Error t value Pr(>|t|)
(Intercept)          316.62      16.16  19.588 4.65e-14 ***
ventilationN2O+O2,op -60.18      22.22  -2.709   0.0139 *
ventilationO2,24h    -38.62      26.06  -1.482   0.1548
---
Signif. codes:  0 '***' 0.001 '**' 0.01 '*' 0.05 '.' 0.1 ' ' 1

Residual standard error: 45.72 on 19 degrees of freedom
Multiple R-squared: 0.2809,    Adjusted R-squared: 0.2052
F-statistic: 3.711 on 2 and 19 DF,   p-value: 0.04359
```

The interpretation of the estimates is that the intercept is the mean in the first group (N20+02, 24h), whereas the two others describe the *difference* between the relevant group and the first one.

There are multiple ways of representing the effect of a factor variable in linear models (and one-way analysis of variance is the simplest example of a linear model with a factor variable). The representations are in terms of *contrasts*, the choice of which can be controlled either by global options or as part of the model formula. We do not go deeply into this but just mention that the contrasts used by default are the so-called *treatment contrasts*, in which the first group is treated as a baseline and the other groups are given relative to that. Concretely, the analysis is performed as a multiple regression analysis (see Chapter 11) by introducing two *dummy variables*, which are 1 for observations in the relevant group and 0 elsewhere.

Among the *t* tests in the table, you can immediately find a test for the hypothesis that the first two groups have the same true mean ($p = 0.0139$) and also whether the first and the third might be identical ($p = 0.1548$). However, a comparison of the last two groups cannot be found. This can be overcome by modifying the factor definition (see the help page for relevel), but that gets tedious when there are more than a few groups.

If we want to compare all groups, we ought to correct for *multiple testing*. Performing many tests will increase the probability of finding one of them to be significant; that is, the *p*-values tend to be exaggerated. A common adjustment method is the *Bonferroni correction*, which is based on the fact that the probability of observing at least one of *n* events is less than the sum of the probabilities for each event. Thus, by dividing the significance level by the number of tests or, equivalently, multiplying the *p*-values, we obtain a *conservative* test where the probability of a significant result is less than or equal to the formal significance level.

A function called pairwise.t.test computes all possible two-group comparisons. It is also capable of making adjustments for multiple comparisons and works like this:

```
> pairwise.t.test(folate, ventilation, p.adj="bonferroni")

        Pairwise comparisons using t tests with pooled SD

data:  folate and ventilation

            N20+02,24h N20+02,op
N20+02,op 0.042        -
02,24h    0.464        1.000

P value adjustment method: bonferroni
```

The output is a table of p-values for the pairwise comparisons. Here, the p-values have been adjusted by the Bonferroni method, where the unadjusted values have been multiplied by the number of comparisons, namely 3. If that results in a value bigger than 1, then the adjustment procedure sets the adjusted p-value to 1.

The default method for `pairwise.t.test` is actually not the Bonferroni correction but a variant due to Holm. In this method, only the smallest p needs to be corrected by the full number of tests, the second smallest is corrected by $n - 1$, etc., unless that would make it smaller than the previous one, since the order of the p-values should be unaffected by the adjustment.

```
> pairwise.t.test(folate,ventilation)

        Pairwise comparisons using t tests with pooled SD

data:   folate and ventilation

            N2O+O2,24h N2O+O2,op
N2O+O2,op 0.042        -
O2,24h    0.310        0.408

P value adjustment method: holm
```

7.1.2   Relaxing the variance assumption

The traditional one-way ANOVA requires an assumption of equal variances for all groups. There is, however, an alternative procedure that does not require that assumption. It is due to Welch and similar to the unequal-variances t test. This has been implemented in the `oneway.test` function:

```
> oneway.test(folate~ventilation)

        One-way analysis of means (not assuming equal variances)

data:   folate and ventilation
F = 2.9704, num df =  2.000, denom df = 11.065, p-value = 0.09277
```

In this case, the p-value increased to a nonsignificant value, presumably related to the fact that the group that seems to differ from the two others also has the largest variance.

It is also possible to perform the pairwise t tests so that they do not use a common pooled standard deviation. This is controlled by the argument `pool.sd`.

```
> pairwise.t.test(folate,ventilation,pool.sd=F)

        Pairwise comparisons using t tests with non-pooled SD

data:   folate and ventilation

          N2O+O2,24h N2O+O2,op
N2O+O2,op 0.087        -
O2,24h    0.321       0.321

P value adjustment method: holm
```

Again, it is seen that the significance disappears as we remove the constraint on the variances.

7.1.3   Graphical presentation

Of course, there are many ways to present grouped data. Here we create a somewhat elaborate plot where the raw data are plotted as a stripchart and overlaid with an indication of means and SEMs (Figure 7.1):

```
> xbar <- tapply(folate, ventilation, mean)
> s <- tapply(folate, ventilation, sd)
> n <- tapply(folate, ventilation, length)
> sem <- s/sqrt(n)
> stripchart(folate~ventilation, method="jitter",
+    jitter=0.05, pch=16, vert=T)
> arrows(1:3,xbar+sem,1:3,xbar-sem,angle=90,code=3,length=.1)
> lines(1:3,xbar,pch=4,type="b",cex=2)
```

Here we used pch=16 (small plotting dots) in stripchart and put vertical=T to make the "strips" vertical.

The error bars have been made with arrows, which adds arrows to a plot. We slightly abuse the fact that the angle of the arrowhead is adjustable to create the little crossbars at either end. The first four arguments specify the endpoints, (x_1, y_1, x_2, y_2); the angle argument gives the angle between the lines of the arrowhead and shaft, here set to 90°; and length is the length of the arrowhead (in inches on a printout). Finally, code=3 means that the arrow should have a head at both ends. Note that the x-coordinates of the stripcharts are simply the group numbers.

The indication of averages and the connecting lines are done with lines, where type="b" (both) means that both points and lines are printed, leaving gaps in the lines to make room for the symbols. pch=4 is a cross, and cex=2 requests that the symbols be drawn in double size.

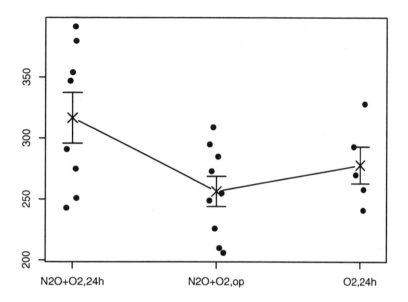

Figure 7.1. "Red cell folate" data with $\bar{x} \pm 1$ SEM.

It is debatable whether you should draw the plot using 1 SEM as is done here or whether perhaps it is better to draw proper confidence intervals for the means (approximately 2 SEM), or maybe even SD instead of SEM. The latter point has to do with whether the plot is to be used in a descriptive or an analytical manner. Standard errors of the mean are not useful for describing the distributions in the groups; they only say how precisely the mean is determined. On the other hand, SDs do not enable the reader to see at a glance which groups are significantly different.

In many fields it appears to have become the tradition to use 1 SEM "because they are the smallest"; that is, it makes differences look more dramatic. Probably, the best thing to do is to follow the traditions in the relevant field and "calibrate your eyeballs" accordingly.

One word of warning, though: At small group sizes, the rule of thumb that the confidence interval is the mean ± 2 SEM becomes badly misleading. At a group size of 2, it actually has to be 12.7 SEM! That is a correction heavily dependent on data having the normal distribution. If you have such small groups, it may be advisable to use a pooled SD for the entire data set rather than the group-specific SDs. This does, of course, require

that you can reasonably assume that the true standard deviation actually
is the same in all groups.

7.1.4   Bartlett's test

Testing whether the distribution of a variable has the same variance in
all groups can be done using Bartlett's test, although like the F test for
comparing two variances, it is rather nonrobust against departures from
the assumption of normal distributions. As in var.test, it is assumed
that the data are from independent groups. The procedure is performed
as follows:

```
> bartlett.test(folate~ventilation)

        Bartlett test of homogeneity of variances

data:   folate by ventilation
Bartlett's K-squared = 2.0951, df = 2, p-value = 0.3508
```

That is, in this case, nothing in the data contradicts the assumption of
equal variances in the three groups.

7.2   Kruskal–Wallis test

A nonparametric counterpart of a one-way analysis of variance is the
Kruskal–Wallis test. As in the Wilcoxon two-sample test (see Section 5.5),
data are replaced with their ranks without regard to the grouping, only
this time the test is based on the between-group sum of squares calculated from the average ranks. Again, the distribution of the test statistic
can be worked out based on the idea that, under the hypothesis of irrelevant grouping, the problem reduces to a combinatorial one of sampling
the within-group ranks from a fixed set of numbers.

You can make R calculate the Kruskal–Wallis test as follows:

```
> kruskal.test(folate~ventilation)

        Kruskal-Wallis rank sum test

data:   folate by ventilation
Kruskal-Wallis chi-squared = 4.1852, df = 2, p-value = 0.1234
```

It is seen that there is no significant difference using this test. This should
not be too surprising in view of the fact that the F test in the one-way analysis of variance was only borderline significant. Also, the Kruskal–Wallis

test is less efficient than its parametric counterpart if the assumptions hold, although it does not invariably give a larger p-value.

7.3 Two-way analysis of variance

One-way analysis of variance deals with one-way classifications of data. It is also possible to analyze data that are cross-classified according to several criteria. When a cross-classified design is *balanced*, then you can almost read the entire statistical analysis from a single analysis of variance table, and that table generally consists of items that are simple to compute, which was very important before the computer era. Balancedness is a concept that is hard to define exactly; for a two-way classification, a sufficient condition is that the cell counts be equal, but there are other balanced designs.

Here we restrict ourselves to the case of a single observation per cell. This typically arises from having multiple measurements on the same experimental unit and in this sense generalizes the paired t test.

Let x_{ij} denote the observation in row i and column j of the $m \times n$ table. This is similar to the notation used for one-way analysis of variance, but notice that there is now a connection between observations with the same j, so that it makes sense to look at both row averages $\bar{x}_{i\cdot}$ and column averages $\bar{x}_{\cdot j}$.

Consequently, it now makes sense to look at both *variation between rows*

$$\mathrm{SSD}_R = n \sum_i (\bar{x}_{i\cdot} - \bar{x}_{\cdot\cdot})^2$$

and *variation between columns*

$$\mathrm{SSD}_C = m \sum_j (\bar{x}_{\cdot j} - \bar{x}_{\cdot\cdot})^2$$

Subtracting these two from the total variation leaves the *residual variation*, which works out as

$$\mathrm{SSD}_{\mathrm{res}} = \sum_i \sum_j (x_{ij} - \bar{x}_{i\cdot} - \bar{x}_{\cdot j} + \bar{x}_{\cdot\cdot})^2$$

This corresponds to a statistical model in which it is assumed that the observations are composed of a general level, a row effect, and a column effect plus a noise term:

$$X_{ij} = \mu + \alpha_i + \beta_j + \epsilon_{ij} \qquad \epsilon_{ij} \sim N(0, \sigma^2)$$

The parameters of this model are not uniquely defined unless we impose some restriction on the parameters. If we impose $\sum \alpha_i = 0$ and $\sum \beta_j = 0$, then the estimates of α_i, β_j, and μ turn out to be $\bar{x}_{i\cdot} - \bar{x}_{\cdot\cdot}$, $\bar{x}_{\cdot j} - \bar{x}_{\cdot\cdot}$, and $\bar{x}_{\cdot\cdot}$.

Dividing the sums of squares by their respective degrees of freedom $m - 1$ for SSD_R, $n - 1$ for SSD_C, and $(m - 1)(n - 1)$ for SSD_{res}, we get a set of mean squares. F tests for no row and column effect can be carried out by dividing the respective mean squares by the residual mean square.

It is important to notice that this works out so nicely only because of the balanced design. If you have a table with "holes" in it, the analysis is considerably more complicated. The simple formulas for the sum of squares are no longer valid and, in particular, the order independence is lost, so that there is no longer a single SSD_C but ones with and without adjusting for row effects.

To perform a two-way ANOVA, it is necessary to have data in one vector, with the two classifying factors parallel to it. We consider an example concerning heart rate after administration of enalaprilate (Altman, 1991, p. 327). Data are found in this form in the heart.rate data set:

```
> attach(heart.rate)
> heart.rate
    hr subj time
1   96    1    0
2  110    2    0
3   89    3    0
4   95    4    0
5  128    5    0
6  100    6    0
7   72    7    0
8   79    8    0
9  100    9    0
10  92    1   30
11 106    2   30
12  86    3   30
13  78    4   30
14 124    5   30
15  98    6   30
16  68    7   30
17  75    8   30
18 106    9   30
19  86    1   60
20 108    2   60
21  85    3   60
22  78    4   60
23 118    5   60
24 100    6   60
25  67    7   60
26  74    8   60
27 104    9   60
```

```
28  92    1  120
29 114    2  120
30  83    3  120
31  83    4  120
32 118    5  120
33  94    6  120
34  71    7  120
35  74    8  120
36 102    9  120
```

If you look inside the heart.rate.R file in the data directory of the ISwR package, you will see that the actual definition of the data frame is

```
heart.rate <- data.frame(hr = c(96,110,89,95,128,100,72,79,100,
                              92,106,86,78,124,98,68,75,106,
                              86,108,85,78,118,100,67,74,104,
                              92,114,83,83,118,94,71,74,102),
                       subj=gl(9,1,36),
                       time=gl(4,9,36,labels=c(0,30,60,120)))
```

The gl (*generate levels*) function is specially designed for generating patterned factors for balanced experimental designs. It has three arguments: the number of levels, the block length (how many times each level should repeat), and the total length of the result. The two patterns in the data frame are thus

```
> gl(9,1,36)
 [1] 1 2 3 4 5 6 7 8 9 1 2 3 4 5 6 7 8 9 1 2 3 4 5 6 7 8 9 1 2 3 4
[32] 5 6 7 8 9
Levels:  1 2 3 4 5 6 7 8 9
> gl(4,9,36,labels=c(0,30,60,120))
 [1] 0    0    0    0    0    0    0    0    0    30   30   30   30   30   30
[16] 30   30   30   60   60   60   60   60   60   60   60   60   120 120 120
[31] 120 120 120 120 120 120
Levels:  0 30 60 120
```

Once the variables have been defined, the two-way analysis of variance is specified simply by

```
> anova(lm(hr~subj+time))
Analysis of Variance Table

Response: hr
          Df Sum Sq Mean Sq F value    Pr(>F)
subj       8 8966.6  1120.8 90.6391 4.863e-16 ***
time       3  151.0    50.3  4.0696   0.01802 *
Residuals 24  296.8    12.4
---
Signif. codes:  0 `***' 0.001 `**' 0.01 `*' 0.05 `.' 0.1 ` ' 1
```

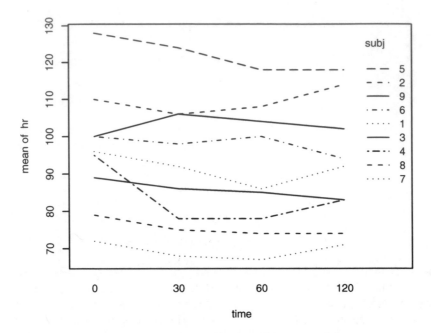

Figure 7.2. Interaction plot of heart-rate data.

Interchanging `subj` and `time` in the model formula (`hr~time+subj`) yields exactly the same analysis except for the order of the rows of the ANOVA table. This is because we are dealing with a balanced design (a complete two-way table with no missing values). In unbalanced cases, the factor order will matter.

7.3.1   Graphics for repeated measurements

At least for your own use, it is useful to plot a "spaghettigram" of the data; that is, a plot where data from the same subject are connected with lines. To this end, you can use the function `interaction.plot`, which graphs the values against one factor while connecting data for the other factor with line segments to form traces.

```
> interaction.plot(time, subj, hr)
```

In fact there is a fourth argument, which specifies what should be done in case, there is more than one observation per cell. By default, the mean is taken, which is the reason why the y-axis in Figure 7.2 reads "mean of hr".

If you prefer to have the values plotted according to the times of measurement (which are not equidistant in this example), you could instead write (resulting plot not shown)

```
> interaction.plot(ordered(time),subj,hr)
```

7.4 The Friedman test

A nonparametric counterpart of two-way analysis of variance exists for the case with one observation per cell. Friedman's test is based on ranking observations *within each row* assuming that if there is no column effect then all orderings should be equally likely. A test statistic based on the column sum of squares can be calculated and normalized to give a χ^2-distributed test statistic.

In the case of two columns, the Friedman test is equivalent to the *sign test,* in which one uses the binomial distribution to test for equal probabilities of positive and negative differences within pairs. This is a rather less sensitive test than the Wilcoxon signed-rank test discussed in Section 5.2.

Practical application of the Friedman test is as follows:

```
> friedman.test(hr~time|subj,data=heart.rate)

        Friedman rank sum test

data:   hr and time and subj
Friedman chi-squared = 8.5059, df = 3, p-value = 0.03664
```

Notice that the blocking factor is specified in a model formula using the vertical bar, which may be read as "time within `subj`". It is seen that the test is not quite as strongly significant as the parametric counterpart. This is unsurprising since the latter test is more powerful when its assumptions are met.

7.5 The ANOVA table in regression analysis

We have seen the use of analysis of variance tables in grouped and cross-classified experimental designs. However, their use is not restricted to these designs but applies to the whole class of *linear models* (more on this in Chapter 12).

The variation between and within groups for a one-way analysis of variance generalizes to *model variation* and *residual variation*

$$SSD_{model} = \sum_i (\hat{y}_i - \bar{y}.)^2$$

$$SSD_{res} = \sum_i (y_i - \hat{y}_i)^2$$

which partition the total variation $\sum_i (y_i - \bar{y}.)^2$. This applies only when the model contains an intercept; see Section 12.2. The role of the group means in the one-way classification is taken over by the fitted values $\hat{y}_i$ in the more general linear model.

An F test for significance of the model is available in direct analogy with Section 7.1. In simple linear regression, this test is equivalent to testing that the regression coefficient is zero.

The analysis of variance table corresponding to a regression analysis can be extracted with the function `anova`, just as for one- and two-way analyses of variance. For the `thuesen` example, it will look like this:

```
> attach(thuesen)
> lm.velo <- lm(short.velocity~blood.glucose)
> anova(lm.velo)
Analysis of Variance Table

Response: short.velocity
              Df  Sum Sq Mean Sq F value Pr(>F)
blood.glucose  1 0.20727 0.20727   4.414 0.0479 *
Residuals     21 0.98610 0.04696
---
Signif. codes:  0 '***' 0.001 '**' 0.01 '*' 0.05 '.' 0.1 ' ' 1
```

Notice that the F test gives the same p-value as the t test for a zero slope from Section 6.1. It is the same F test that gets printed at the end of the `summary` output:

```
...
Residual standard error: 0.2167 on 21 degrees of freedom
Multiple R-Squared: 0.1737,     Adjusted R-squared: 0.1343
F-statistic: 4.414 on 1 and 21 DF, p-value: 0.0479
```

The remaining elements of the three output lines above may also be derived from the ANOVA table. "Residual standard error" is the square root of "Residual mean squares", namely $0.2167 = \sqrt{0.04696}$. R^2 is the proportion of the total sum of squares explained by the regression line, $0.1737 = 0.2073/(0.2073 + 0.9861)$; and, finally, the adjusted R^2 is the relative improvement of the residual variance, $0.1343 = (v - 0.04696)/v$, where $v = (0.2073 + 0.9861)/22 = 0.05425$ is the variance of `short.velocity` if the glucose values are not taken into account.

7.6   Exercises

7.1   The `zelazo` data are in the form of a list of vectors, one for each of the four groups. Convert the data to a form suitable for the use of `lm`, and calculate the relevant test. Consider *t* tests comparing selected subgroups or obtained by combining groups.

7.2   In the `lung` data, do the three measurement methods give systematically different results? If so, which ones appear to be different?

7.3   Repeat the previous exercises using the `zelazo` and `lung` data with the relevant nonparametric tests.

7.4   The `igf1` variable in the `juul` data set is arguably skewed and has different variances across Tanner groups. Try to compensate for this using logarithmic and square-root transformations, and use the Welch test. However, the analysis is still problematic — why?

8

Tabular data

This chapter describes a series of functions designed to analyze tabular data. Specifically, we look at the functions `prop.test`, `binom.test`, `chisq.test`, and `fisher.test`.

8.1 Single proportions

Tests of single proportions are generally based on the binomial distribution (see Section 3.3) with size parameter N and probability parameter p. For large sample sizes, this can be well approximated by a normal distribution with mean Np and variance $Np(1 - p)$. As a rule of thumb, the approximation is satisfactory when the expected numbers of "successes" and "failures" are both larger than 5.

Denoting the observed number of "successes" by x, the test for the hypothesis that $p = p_0$ can be based on

$$u = \frac{x - Np_0}{\sqrt{Np_0(1 - p_0)}}$$

which has an approximate normal distribution with mean zero and standard deviation 1, or on u^2, which has an approximate χ^2 distribution with 1 degree of freedom.

P. Dalgaard, *Introductory Statistics with R*,
DOI: 10.1007/978-0-387-79054-1_8, © Springer Science+Business Media, LLC 2008

The normal approximation can be somewhat improved by the *Yates correction*, which shrinks the observed value by half a unit towards the expected value when calculating u.

We consider an example (Altman, 1991, p. 230) where 39 of 215 randomly chosen patients are observed to have asthma and one wants to test the hypothesis that the probability of a "random patient" having asthma is 0.15. This can be done using prop.test:

```
> prop.test(39,215,.15)

        1-sample proportions test with continuity correction

data:   39 out of 215, null probability 0.15
X-squared = 1.425, df = 1, p-value = 0.2326
alternative hypothesis: true p is not equal to 0.15
95 percent confidence interval:
 0.1335937 0.2408799
sample estimates:
        p
0.1813953
```

The three arguments to prop.test are the number of positive outcomes, the total number, and the (theoretical) probability parameter that you want to test for. The latter is 0.5 by default, which makes sense for symmetrical problems, but this is not the case here. The amount 15% is a bit synthetic since it is rarely the case that one has a specific a priori value to test for. It is usually more interesting to compute a confidence interval for the probability parameter, such as is given in the last part of the output. Notice that we have a slightly unfortunate double usage of the symbol p as the probability parameter of the binomial distribution and as the test probability or p-value.

You can also use binom.test to obtain a test in the binomial distribution. In that way, you get an exact test probability, so it is generally preferable to using prop.test, but prop.test can do more than testing single proportions. The procedure to obtain the p-value is to calculate the point probabilities for all the possible values of x and sum those that are less than or equal to the point probability of the observed x.

```
> binom.test(39,215,.15)

        Exact binomial test

data:   39 and 215
number of successes = 39, number of trials = 215, p-value = 0.2135
alternative hypothesis: true probability  ...  not equal to 0.15
95 percent confidence interval:
 0.1322842 0.2395223
sample estimates:
```

```
probability of success
            0.1813953
```

The "exact" confidence intervals at the 0.05 level are actually constructed from the two one-sided tests at the 0.025 level. Finding an exact confidence interval using two-sided tests is not a well-defined problem (see Exercise 8.5).

8.2 Two independent proportions

The function `prop.test` can also be used to compare two or more proportions. For that purpose, the arguments should be given as two vectors, where the first contains the number of positive outcomes and the second the total number for each group.

The theory is similar to that for a single proportion. Consider the difference in the two proportions $d = x_1/N_1 - x_2/N_2$, which will be approximately normally distributed with mean zero and variance $V_p(d) = (1/N_1 + 1/N_2) \times p(1 - p)$ if the counts are binomially distributed with the same p parameter. So to test the hypothesis that $p_1 = p_2$, plug the common estimate $\hat{p} = (x_1 + x_2)/(n_1 + n_2)$ into the variance formula and look at $u = d/\sqrt{V_{\hat{p}}(d)}$, which approximately follows a standard normal distribution, or look at u^2, which is approximately $\chi^2(1)$-distributed. A Yates-type correction is possible, but we skip the details.

For illustration, we use an example originally due to Lewitt and Machin (Altman, 1991, p. 232):

```
> lewitt.machin.success <- c(9,4)
> lewitt.machin.total <- c(12,13)
> prop.test(lewitt.machin.success,lewitt.machin.total)

        2-sample test for equality of proportions with continuity
        correction

data:  lewitt.machin.success out of lewitt.machin.total
X-squared = 3.2793, df = 1, p-value = 0.07016
alternative hypothesis: two.sided
95 percent confidence interval:
 0.01151032 0.87310506
sample estimates:
   prop 1    prop 2
0.7500000 0.3076923
```

The confidence interval given is for the *difference* in proportions. The theory behind its calculation is similar to that of the test, but there are some technical complications, and a different approximation is used.

You can also perform the test without the Yates continuity correction. This is done by adding the argument `correct=F`. The continuity correction makes the confidence interval somewhat wider than it would otherwise be, but notice that it nevertheless does not contain zero. Thus, the confidence interval is contradicting the test, which says that there is *no* significant difference between the two groups with a two-sided test. The explanation lies in the different approximations, which becomes important for tables as sparse as the present one.

If you want to be sure that at least the *p*-value is correct, you can use Fisher's exact test. We illustrate this using the same data as in the preceding section. The test works by making the calculations in the conditional distribution of the 2×2 table given both the row and column marginals. This can be difficult to envision, but think of it like this: Take 13 white balls and 12 black balls (success and failure, respectively), and sample the balls without replacement into two groups of sizes 12 and 13. The number of white balls in the first group obviously defines the whole table, and the point is that its distribution can be found as a purely combinatorial problem. The distribution is known as the *hypergeometric distribution*.

The relevant function is `fisher.test`, which requires that data be given in matrix form. This is obtained as follows:

```
> matrix(c(9,4,3,9),2)
     [,1] [,2]
[1,]    9    3
[2,]    4    9

> lewitt.machin <- matrix(c(9,4,3,9),2)
> fisher.test(lewitt.machin)

        Fisher's Exact Test for Count Data

data:  lewitt.machin
p-value = 0.04718
alternative hypothesis: true odds ratio is not equal to 1
95 percent confidence interval:
  0.9006803 57.2549701
sample estimates:
odds ratio
  6.180528
```

Notice that the second column of the table needs to be the number of negative outcomes, not the total number of observations.

Notice also that the confidence interval is for the *odds ratio*; that is, for the estimate of $(p_1/(1-p_1))/(p_2/(1-p_2))$. One can show that if the *p*s are not identical, then the conditional distribution of the table depends only on the odds ratio, so it is the natural measure of association in connection with the Fisher test. The exact distribution of the test can be worked out also when the odds ratio differs from 1, but there is the same complication as with `binom.test` that a two-sided 95% confidence interval must be pasted together from two one-sided 97.5% intervals. This leads to the opposite inconsistency as with `prop.test`: The test is (barely) significant, but the confidence interval for the odds ratio includes 1.

The standard χ^2 test (see also Section 8.4) in `chisq.test` works with data in matrix form, like `fisher.test` does. For a 2 × 2 table, the test is exactly equivalent to `prop.test`.

```
> chisq.test(lewitt.machin)

        Pearson's Chi-squared test with Yates' continuity
        correction

data:   lewitt.machin
X-squared = 3.2793, df = 1, p-value = 0.07016
```

8.3   *k* proportions, test for trend

Sometimes you want to compare more than two proportions. In that case, the categories are often ordered so that you would expect to find a decreasing or increasing trend in the proportions with the group number.

The example used in this section concerns data from a group of women giving birth where it was recorded whether the child was delivered by caesarean section and what shoe size the mother used (Altman, 1991, p. 229).

The table looks like this:

```
> caesar.shoe
      <4   4 4.5   5 5.5   6+
Yes    5   7   6   7   8   10
No    17  28  36  41  46  140
```

To compare $k > 2$ proportions, another test based on the normal approximation is available. It consists of the calculation of a weighted sum of squared deviations between the observed proportions in each group and the overall proportion for all groups. The test statistic has an approximate χ^2 distribution with $k - 1$ degrees of freedom.

To use `prop.test` on a table like `caesar.shoe`, we need to convert it to a vector of "successes" (which in this case is close to being the opposite) and a vector of "trials". The two vectors can be computed like this:

```
> caesar.shoe.yes <- caesar.shoe["Yes",]
> caesar.shoe.total <- margin.table(caesar.shoe,2)
> caesar.shoe.yes
  <4   4 4.5   5 5.5   6+
   5   7   6   7   8   10
> caesar.shoe.total
  <4   4 4.5   5 5.5   6+
  22  35  42  48  54  150
```

Thereafter it is easy to perform the test:

```
> prop.test(caesar.shoe.yes,caesar.shoe.total)
        6-sample test for equality of proportions without
        continuity correction

data:   caesar.shoe.yes out of caesar.shoe.total
X-squared = 9.2874, df = 5, p-value = 0.09814
alternative hypothesis: two.sided
sample estimates:
     prop 1     prop 2     prop 3     prop 4     prop 5     prop 6
0.22727273 0.20000000 0.14285714 0.14583333 0.14814815 0.06666667

Warning message:
In prop.test(caesar.shoe.yes, caesar.shoe.total) :
  Chi-squared approximation may be incorrect
```

It is seen that the test comes out nonsignificant, but the subdivision is really unreasonably fine in view of the small number of caesarean sections. Notice, by the way, the warning about the χ^2 approximation being dubious, which is prompted by some cells having an expected count less than 5.

You can test for a trend in the proportions using `prop.trend.test`. It takes three arguments: x, n, and `score`. The first two of these are exactly as in `prop.test`, whereas the last one is the score given to the groups, by default simply $1, 2, \ldots, k$. The basis of the test is essentially a weighted linear regression of the proportions on the group scores, where we test for a zero slope, which becomes a χ^2 test on 1 degree of freedom.

```
> prop.trend.test(caesar.shoe.yes,caesar.shoe.total)

        Chi-squared Test for Trend in Proportions

data:   caesar.shoe.yes out of caesar.shoe.total ,
  using scores: 1 2 3 4 5 6
X-squared = 8.0237, df = 1, p-value = 0.004617
```

So if we assume that the effect of shoe size is linear in the group score, *then* we can see a significant difference. This kind of assumption should not be thought of as something that must hold for the test to be valid. Rather, it indicates the rough type of alternative to which the test should be sensitive.

The effect of using a trend test can be viewed as an approximate subdivision of the test for equal proportions ($\chi^2 = 9.29$) into a contribution from the linear effect ($\chi^2 = 8.02$) on 1 degree of freedom and a contribution from deviations from the linear trend ($\chi^2 = 1.27$) on 4 degrees of freedom. So you could say that the test for equal proportions is being diluted or wastes degrees of freedom on testing for deviations in a direction we are not really interested in.

8.4 $r \times c$ tables

For the analysis of tables with more than two classes on both sides, you can use `chisq.test` or `fisher.test`, although you should note that the latter can be very computationally demanding if the cell counts are large and there are more than two rows or columns. We have already seen `chisq.test` in a simple example, but with larger tables, some additional features are of interest.

An $r \times c$ table looks like this:

n_{11}	n_{12}	$\cdots$	n_{1c}	$n_{1\cdot}$
n_{21}	n_{22}	$\cdots$	n_{2c}	$n_{2\cdot}$
$\vdots$	$\vdots$		$\vdots$	$\vdots$
n_{r1}	n_{r2}	$\cdots$	n_{rc}	$n_{r\cdot}$
$n_{\cdot 1}$	$n_{\cdot 2}$	$\cdots$	$n_{\cdot c}$	$n_{\cdot\cdot}$

Such a table can arise from several different sampling plans, and the notion of "no relation between rows and columns" is correspondingly different. The total in each row might be fixed in advance, and you would be interested in testing whether the distribution over columns is the same for each row, or vice versa if the column totals were fixed. It might also be the case that only the total number is chosen and the individuals are grouped randomly according to the row and column criteria. In the latter case, you would be interested in testing the hypothesis of *statistical independence*, that the probability of an individual falling into the ijth cell is the product $p_{i\cdot} p_{\cdot j}$ of the marginal probabilities. However, the analysis of the table turns out to be the same in all cases.

If there is no relation between rows and columns, then you would expect to have the following cell values:

$$E_{ij} = \frac{n_{i\cdot} \times n_{\cdot j}}{n_{\cdot\cdot}}$$

This can be interpreted as distributing each row total according to the proportions in each column (or vice versa) or as distributing the grand total according to the products of the row and column proportions.

The test statistic

$$X^2 = \sum \frac{(O - E)^2}{E}$$

has an approximate χ^2 distribution with $(r - 1) \times (c - 1)$ degrees of freedom. Here the sum is over the entire table and the ij indices have been omitted. O denotes the observed values and E the expected values as described above.

We consider the table with caffeine consumption and marital status from Section 4.5 and compute the χ^2 test:

```
> caff.marital <- matrix(c(652,1537,598,242,36,46,38,21,218
+ ,327,106,67),
+ nrow=3,byrow=T)
> colnames(caff.marital) <- c("0","1-150","151-300",">300")
> rownames(caff.marital) <- c("Married","Prev.married","Single")
> caff.marital
                0 1-150 151-300 >300
Married       652  1537     598  242
Prev.married   36    46      38   21
Single        218   327     106   67
> chisq.test(caff.marital)

        Pearson's Chi-squared test

data:  caff.marital
X-squared = 51.6556, df = 6, p-value = 2.187e-09
```

The test is highly significant, so we can safely conclude that the data contradict the hypothesis of independence. However, you would generally also like to know the nature of the deviations. To that end, you can look at some extra components of the return value of chisq.test.

Notice that chisq.test (just like lm) actually returns more information than what is commonly printed:

```
> chisq.test(caff.marital)$expected
                 0      1-150    151-300       >300
Married    705.83179 1488.01183 578.06533 257.09105
Prev.married 32.85648   69.26698  26.90895  11.96759
Single     167.31173  352.72119 137.02572  60.94136
> chisq.test(caff.marital)$observed
             0 1-150 151-300 >300
Married    652  1537     598  242
Prev.married 36    46      38   21
Single     218   327     106   67
```

These two tables may then be scrutinized to see where the differences lie. It is often useful to look at a table of the contributions from each cell to the total χ^2. Such a table cannot be directly extracted, but it is easy to calculate:

```
> E <- chisq.test(caff.marital)$expected
> O <- chisq.test(caff.marital)$observed
> (O-E)^2/E
                 0     1-150   151-300      >300
Married    4.1055981 1.612783 0.6874502 0.8858331
Prev.married 0.3007537 7.815444 4.5713926 6.8171090
Single    15.3563704 1.875645 7.0249243 0.6023355
```

There are some large contributions, particularly from too many "abstaining" singles, and the distribution among previously married is shifted in the direction of a larger intake — insofar as they consume caffeine at all. Still, it is not easy to find a simple description of the deviation from independence in these data.

You can also use `chisq.test` directly on raw (untabulated) data, here using the `juul` data set from Section 4.5:

```
> attach(juul)
> chisq.test(tanner,sex)

        Pearson's Chi-squared test

data:  tanner and sex
X-squared = 28.8672, df = 4, p-value = 8.318e-06
```

It may not really be relevant to test for independence between these particular variables. The definition of Tanner stages is gender-dependent by nature.

8.5   Exercises

8.1   Reconsider the situation of Exercise 3.3, where 10 consecutive patients had operations without complications and the expected rate was

20%. Calculate the relevant one-sided test in the binomial distribution. How large a sample (still with zero complications) would be necessary to obtain statistical significance?

8.2   In 747 cases of "Rocky Mountain spotted fever" from the western United States, 210 patients died. Out of 661 cases from the eastern United States, 122 died. Is the difference statistically significant? (See also Exercise 13.4.)

8.3   Two drugs for the treatment of peptic ulcer were compared (Campbell and Machin, 1993, p. 72). The results were as follows:

	Healed	Not Healed	Total
Pirenzepine	23	7	30
Trithiozine	18	13	31
Total	41	20	61

Compute the χ^2 test and Fisher's exact test and discuss the difference. Find an approximate 95% confidence interval for the difference in healing probability.

8.4   (From "Mathematics 5" exam, University of Copenhagen, Summer 1969.) From September 20, 1968, to February 1, 1969, an instructor consumed 254 eggs. Every day, he recorded how many eggs broke during boiling so that the white ran out and how many cracked so that the white did not run out. Additionally, he recorded whether the eggs were size A or size B. From February 4, 1969, until April 10, 1969, he consumed 130 eggs, but this time he used a "piercer" to create a small hole in the egg to prevent breaking and cracking. The results were as follows:

Period	Size	Total	Broken	Cracked
Sept. 20–Feb. 1	A	54	4	8
Sept. 20–Feb. 1	B	200	15	28
Feb. 4–Apr. 10	A	60	4	9
Feb. 4–Apr. 10	B	70	1	7

Investigate whether or not the piercer seems to have had an effect.

8.5   Make a plot of the two-sided p-value for testing that the probability parameter is x when the observations are 3 successes in 15 trials for x varying from 0 to 1 in steps of 0.001. Explain what makes the definition of a two-sided confidence interval difficult.

9

Power and the computation of sample size

A statistical test will not be able to detect a true difference if the sample size is too small compared with the magnitude of the difference. When designing experiments, the experimenter should try to ensure that a sufficient amount of data are collected to be reasonably sure that a difference of a specified size will be detected. R has methods for doing these calculations in the simple cases of comparing means using one- or two-sample t tests and comparing two proportions.

9.1 The principles of power calculations

This section outlines the theory of power calculations and sample-size choice. If you are practically inclined and just need to find the necessary sample size in a particular situation, you can safely skim this section and move quickly to subsequent sections that contain the actual R calls.

The basic idea of a hypothesis test should be clear by now. A test statistic is defined, and its value is used to decide whether or not you can accept the (null) hypothesis. Acceptance and rejection regions are set up so that the probability of getting a test statistic that falls into the rejection region is a specified significance level (α) if the null hypothesis is true. In the present context, it is useful to stick to this formulation (as opposed to the use of p-values), as rigid as it might be.

P. Dalgaard, *Introductory Statistics with R*,
DOI: 10.1007/978-0-387-79054-1_9, © Springer Science+Business Media, LLC 2008

Since data are sampled at random, there is always a risk of reaching a wrong conclusion, and things can go wrong in two ways:

- The hypothesis is correct, but the test rejects it (type I error).
- The hypothesis is wrong, but the test accepts it (type II error).

The risk of a type I error is the significance level. The risk of a type II error will depend on the size and nature of the deviation you are trying to detect. If there is very little difference, then you do not have much of a chance of detecting it. For this reason, some statisticians disapprove of terms like "acceptance region" because you can never prove that there is no difference — you can only fail to prove that there is one.

The probability of rejecting a false hypothesis is called the *power* of the test, and methods exist for calculating or approximating the power in the most important practical situations. It is inconvenient to talk further about these matters in the abstract, so let us move on to some concrete examples.

9.1.1 Power of one-sample and paired t tests

Consider the case of the comparison of a sample mean to a given value. For example, in a matched trial we wish to test whether the difference between treatment A and treatment B is zero using a paired *t* test (described in Chapter 5).

We call the true difference δ. Even if the null hypothesis is not true, we can still work out the distribution of the test statistic, provided the other model assumptions hold. It is called the *noncentral t distribution* and depends on a noncentrality parameter as well as the usual degrees of freedom. For the paired *t* test, the noncentrality parameter v is a function of δ, the standard deviation of differences σ, and the sample size n and equals

$$v = \frac{\delta}{\sigma/\sqrt{n}}$$

That is, it is simply the true difference divided by the standard error of the mean.

The cumulative noncentral *t* distribution is available in R simply by adding an ncp argument to the pt function. Figure 9.1 shows a plot of pt with ncp=3 and df=25. A vertical line indicates the upper end of the acceptance region for a two-sided test at the 0.05 significance level. The plot was created as follows:

```
> curve(pt(x,25,ncp=3), from=0, to=6)
> abline(v=qt(.975,25))
```

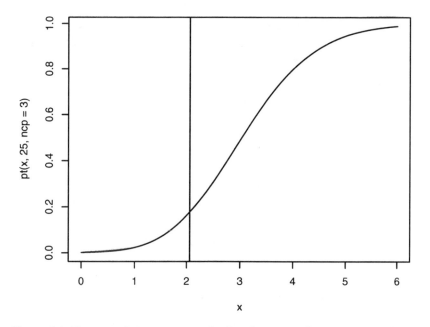

Figure 9.1. The cumulative noncentral t distribution with $\nu = 3$ and 25 degrees of freedom. The vertical line marks the upper significance limit for a two-sided test at the 0.05 level.

The plot shows the main part of the distribution falling in the rejection region. The probability of getting a value in the acceptance region can be seen from the graph as the intersection between the curve and the vertical line. (Almost! See Exercise 9.4.) This value is easily calculated as

```
> pt (qt (.975,25),25,ncp=3)
[1] 0.1779891
```

or roughly 0.18. The power of the test is the opposite, the probability of getting a significant result. In this case it is 0.82, and it is of course desirable to have the power as close to 1 as possible.

Notice that the power (traditionally denoted β) depends on four quantities: δ, σ, n, and α. If we fix any three of these, we can adjust the fourth to achieve a given power. This can be used to determine the necessary sample size for an experiment: You need to specify a desired power ($\beta = 0.80$ and $\beta = 0.90$ are common choices), the significance level (usually given by convention as $\alpha = 0.05$), a guess of the standard deviation, and δ, which is known as the "minimal relevant difference" (MIREDIF) or "smallest meaningful difference" (SMD). This gives an equation that you can solve for n. The result will generally be a fractional number, which should of course be rounded up.

You can also work on the opposite problem and answer the following question: Given a feasible sample size, how large a difference should you reasonably be able to detect?

Sometimes a shortcut is made by expressing δ relative to the standard deviation, in which case you would simply set σ to 1.

9.1.2   Power of two-sample t test

Procedures for two-sample t tests are essentially the same as for the one-sample case, except for the calculation of the noncentrality parameter, which is calculated as

$$\nu = \frac{\delta}{\sigma\sqrt{1/n_1 + 1/n_2}}$$

It is generally assumed that the variance is the same in the two groups; that is, using the Welch procedure is not considered. In sample-size calculations, one usually assumes that the group sizes are the same since that gives the optimal power for a given total number of observations.

9.1.3   Approximate methods

For hand calculations, the power calculations can be considerably simplified by assuming that the standard deviation is known, so that the t test is replaced by a test in the standard normal distribution. The practical advantage is that the approximate formula for the power is easily inverted to give an explicit formula for n. For the one- and two-sample cases, this works out as

$$n = \left(\frac{\Phi_{\alpha/2} + \Phi_{\beta}}{\delta/\sigma}\right)^2 \qquad \text{one-sample}$$

$$n = 2 \times \left(\frac{\Phi_{\alpha/2} + \Phi_{\beta}}{\delta/\sigma}\right)^2 \qquad \text{two-sample, each group}$$

with the Φ_x denoting quantiles on the normal distribution. This is for two-sided tests. For one-sided tests, use α instead of $\alpha/2$.

These formulas are often found in textbooks, and some computer programs implement them rather than the more accurate method described earlier. They do have the advantage of more clearly displaying theoretical properties such as the proportionality of δ and $1/\sqrt{n}$ for a given power.

However, they become numerically unsatisfactory when the degrees of freedom falls below 20 or so.

9.1.4 Power of comparisons of proportions

Suppose you wish to compare the morbidity between two populations and have to decide the number of persons to sample from each population. That is, you plan to perform a comparison of two binomial distributions as in Section 8.2 using prop.test or chisq.test.

For binomial comparisons, exact power calculations become unwieldy, so we rely on normal approximations to the binomial distribution. The power will depend on the probabilities in both groups, not just their difference. As for the t test, the group sizes are assumed to be equal. The theoretical derivation of the power proceeds along the same lines as before by calculating the distribution of $\hat{p}_1 - \hat{p}_2$ when $p_1 \neq p_2$ and the probability that it falls outside the range of values compatible with the hypothesis $p_1 = p_2$. Assuming equal numbers in the two groups, this leads to the sample-size formula

$$n = \left(\frac{\Phi_{\alpha/2}\sqrt{2p(1-p)} + \Phi_\beta\sqrt{p_1(1-p_1) + p_2(1-p_2)}}{|p_2 - p_1|} \right)^2$$

in which $p = (p_1 + p_2)/2$.

Since the method is only approximate, the results are not reliable unless the expected number in each of the four cells in the 2×2 table is greater than 5.

9.2 Two-sample problems

The following example is from Altman (1991, p. 457) and concerns the influence of milk on growth. Two groups are to be given different diets, and their growth will be measured. We wish to compute the sample size that with a power of 90%, using a two-sided test at the 1% level, can find a difference of 0.5 cm in a distribution with a standard deviation of 2 cm. This is done as follows:

```
> power.t.test(delta=0.5, sd=2, sig.level = 0.01, power=0.9)

     Two-sample t test power calculation

          n = 477.8021
      delta = 0.5
```

```
            sd = 2
     sig.level = 0.01
         power = 0.9
   alternative = two.sided
```

```
NOTE: n is number in *each* group
```

delta stands for the "true difference", and sd is the standard deviation. As is seen, the calculation may return a fractional number of experimental units. This would, of course, in practice be rounded up to 478. In the original reference, a method employing nomograms (a graphical technique) is used and the value obtained is 450. The difference is probably due to difficulty in reading the value off the nomogram scale. To know which power you would actually obtain with 450 in each group, you would enter

```
> power.t.test (n=450, delta=0.5, sd=2, sig.level = 0.01)

      Two-sample t test power calculation

             n = 450
         delta = 0.5
            sd = 2
     sig.level = 0.01
         power = 0.8784433
   alternative = two.sided
```

```
NOTE: n is number in *each* group
```

The system is that exactly four out of five arguments (power, sig.level, delta, sd, and n) are given, and the function computes the missing one (defaults exist to set sd=1 and sig.level=0.05 — if you wish to have those calculated, explicitly pass them as NULL). In addition, there are two optional arguments: alternative, which can be used to specify one-sided tests; and type, which can be used to specify that you want to handle a one-sample problem. An example of the former is

```
> power.t.test (delta=0.5, sd=2, sig.level = 0.01, power=0.9,
+ alt="one.sided")

      Two-sample t test power calculation

             n = 417.898
         delta = 0.5
            sd = 2
     sig.level = 0.01
         power = 0.9
   alternative = one.sided
```

```
NOTE: n is number in *each* group
```

9.3   One-sample problems and paired tests

One-sample problems are handled by adding `type="one.sample"` in the call to `power.t.test`. Similarly, paired tests are specified with `type="paired"`; although these reduce to one-sample tests by forming differences, the printout will be slightly different.

One pitfall when planning a study with paired data is that the literature sometimes gives the intra-individual variation as "standard deviation of repeated measurements on the same person" or similar. These may be calculated by measuring a number of persons several times and computing a common standard deviation within persons. This needs to be multiplied by $\sqrt{2}$ to get the standard deviation of differences, which `power.t.test` requires for paired data. If, for instance, it is known that the standard deviation within persons is about 10, and you want to use a paired test at the 5% level to detect a difference of 10 with a power of 85%, then you should enter

```
> power.t.test(delta=10, sd=10*sqrt(2), power=0.85, type="paired")

        Paired t test power calculation

              n = 19.96892
          delta = 10
             sd = 14.14214
      sig.level = 0.05
          power = 0.85
    alternative = two.sided

  NOTE: n is number of *pairs*, sd is std.dev. of
        *differences* within pairs
```

Notice that `sig.level=0.05` was taken as the default.

9.4   Comparison of proportions

To calculate sample sizes and related quantities for comparisons of proportions, you should use `power.prop.test`. This is based on approximations with the normal distribution, so do not trust the results if any of the expected cell counts drop below 5.

The use of `power.prop.test` is analogous to `power.t.test`, although `delta` and `sd` are replaced by the hypothesized probabilities in the two groups, p1 and p2. Currently, it is not possible to specify that one wants to consider a one-sample problem.

An example is given in Altman (1991, p. 459) in which two groups are administered or not administered nicotine chewing gum and the binary outcome is smoking cessation. The stipulated values are $p_1 = 0.15$ and $p_2 = 0.30$. We want a power of 85%, and the significance level is the traditional 5%. Inserting these values yields

```
> power.prop.test(power=.85,p1=.15,p2=.30)

     Two-sample comparison of proportions power calculation

              n = 137.6040
             p1 = 0.15
             p2 = 0.3
      sig.level = 0.05
          power = 0.85
    alternative = two.sided

NOTE: n is number in *each* group
```

9.5   Exercises

9.1   The ashina trial was designed to have 80% power if the true treatment difference was 15% and the standard deviation of differences within a person was 20%. Comment on the sample size chosen. (The power calculation was originally done using the approximative formula. The imbalance between the group sizes is due to the use of an open randomization procedure.)

9.2   In a trial comparing a binary outcome between two groups, find the required number of patients to find an increase in the success rate from 60% to 75% with a power of 90%. What happens if we reduce the power requirement to 80%?

9.3   Plot the density of the noncentral t distribution for ncp=3 and df=25 and compare it with the distribution of $t + 3$, where t has a central t distribution with df=25.

9.4   In two-sided tests, there is also a risk of falling into the rejection region on the opposite side of the true value. The power calculations in R only take this into account if you set strict=TRUE. Discuss the consequences.

9.5   It is occasionally suggested to choose n to "make the true difference significant". What power would result from choosing n by such a procedure?

10

Advanced data handling

In the preceding text, we have covered a basic set of elementary statistical procedures. In the chapters that follow, we begin to discuss more elaborate statistical modelling.

This is also a natural point to discuss some data handling techniques that are useful in the practical analysis of data but were too advanced to cover in the first two chapters of the book.

10.1 Recoding variables

This section describes some techniques that are used to construct derived variables: grouping quantitative data, combining and renaming factor levels, and handling date values.

10.1.1 The cut function

You may need to convert a quantitative variable to a grouping factor. For instance, you may wish to present your data in terms of age in 5-year groups, but age is in the data set as a quantitative variable, recorded as whole years or perhaps to a finer resolution. This is what the cut function

P. Dalgaard, *Introductory Statistics with R*,
DOI: 10.1007/978-0-387-79054-1_10, © Springer Science+Business Media, LLC 2008

is for. The basic principles are quite simple, although there are some fine points to be aware of.

The function has two basic arguments: a numeric vector and a vector of *breakpoints*. The latter defines a set of intervals into which the variable is grouped. You have to specify both ends of all intervals; that is, the total number of break points must be one more than the number of intervals. It is a common mistake to believe that the outer breakpoints can be omitted but the result for a value outside all intervals is set to NA. The outer breakpoints can be chosen as -Inf and Inf, though.

The intervals are left-open, right-closed by default. That is, they include the breakpoint at the right end of each interval. The lowest breakpoint is *not* included unless you set include.lowest=TRUE, making the first interval closed at both ends.

In (e.g.) epidemiology, you are more likely to want groupings like "40–49 years of age". This opposite convention can be obtained by setting right=FALSE.

Of course, as you switch to left-closed, right-open intervals, the issue of losing the extreme interval endpoint shifts to the other end of the scale. In that case, include.lowest actually includes the *highest* value! In the example below, the difference lies in the inclusion of two subjects who were exactly 16 years old.

```
> age <- subset(juul, age >= 10 & age <= 16)$age
> range(age)
[1] 10.01 16.00
> agegr <- cut(age, seq(10,16,2), right=F, include.lowest=T)
> length(age)
[1] 502
> table(agegr)
agegr
[10,12) [12,14) [14,16]
    190     168     144
> agegr2 <- cut(age, seq(10,16,2), right=F)
> table(agegr2)
agegr2
[10,12) [12,14) [14,16)
    190     168     142
```

It is sometimes desired to split data into roughly equal-sized groups. This can be achieved by using breakpoints computed by quantile, which was described in Section 4.1. For instance, you could do

```
> q <- quantile(age, c(0, .25, .50, .75, 1))
> q
      0%      25%      50%      75%     100%
10.0100 11.3825 12.6400 14.2275 16.0000
```

```
> ageQ <- cut(age, q, include.lowest=T)
> table(ageQ)
ageQ
  [10,11.4]  (11.4,12.6]  (12.6,14.2]   (14.2,16]
        126          125          125         126
```

The level names resulting from cut turn out rather ugly at times. Fortunately they are easily changed. You can modify each of the factors created above as follows:

```
> levels(ageQ) <- c("1st", "2nd", "3rd", "4th")
> levels(agegr) <- c("10-11", "12-13", "14-15")
```

Frank Harrell's Hmisc package contains the cut2 function, which simplifies some of these matters.

10.1.2 Manipulating factor levels

In Section 1.2.8, we used levels(f)<- .... to change the level set of a factor. Some related tasks will be discussed in this section.

First, notice that the conversion from numeric input and renaming of levels can be done in one operation:

```
> pain <- c(0,3,2,2,1)
> fpain <- factor(pain,levels=0:3,
+        labels=c("none","mild","medium","severe"))
```

Beware the slightly confusing distinction between levels and labels. The latter end up being the levels of the result, whereas the former refers to the coding of the input vector (pain in this case). That is, levels refers to the input and labels to the output.

If you do not specify a levels argument, the levels will be the sorted, unique values represented in the vector. This is not always desirable when dealing with text variables since the sorting is *alphabetical*. Consider, for instance,

```
> text.pain <-  c("none","severe", "medium", "medium", "mild")
> factor(text.pain)
[1] none   severe medium medium mild
Levels:  medium mild none severe
```

Another reason for specifying levels is that the default levels, obviously, do not include values that are not present in data. This may or may not be a problem, but it has consequences for later analyses; for instance, whether tables contain zero entries or whether barplots leave space for the empty columns.

The `factor` function works on factors as if they were character vectors, so you can reorder the levels as follows

```
> ftpain <- factor(text.pain)
> ftpain2 <- factor(ftpain,
+                    levels=c("none", "mild", "medium", "severe"))
```

Another typical task is to combine two or more levels. This is often done when groups would otherwise be too small for valid statistical analysis. Say you wish to combine the levels "medium" and "mild" into a single "intermediate" level. For this purpose, the assignment form of `levels` allows the right-hand side to be a list:

```
> ftpain3 <- ftpain2
> levels(ftpain3) <- list(
+           none="none",
+           intermediate=c("mild","medium"),
+           severe="severe")
> ftpain3
[1] none           severe         intermediate intermediate
[5] intermediate
Levels: none intermediate severe
```

However, it is often easier just to change the level names and give the same name to several groups:

```
> ftpain4 <- ftpain2
> levels(ftpain4) <- c("none","intermediate","intermediate","severe")
> ftpain4
[1] none           severe         intermediate intermediate
[5] intermediate
Levels: none intermediate severe
```

The latter method is not quite as general as the former, though. It gives less control over the final ordering of levels.

10.1.3 Working with dates

In epidemiology and survival data, you often deal with time in the form of dates in calendar format. Different formats are used in different places of the world, and the files you have to read were not necessarily written in the same region as the one you are currently in. The "Date" class and associated conversion routines exist to help you deal with the complexity.

As an example, consider the Estonian stroke study, a preprocessed version of which is contained in the data frame `stroke`. The raw data files can be found in the `rawdata` directory of the `ISwR` package and read using the following code:

```
> stroke <- read.csv2(
+    system.file("rawdata","stroke.csv", package="ISwR"),
+    na.strings=".")
> names(stroke) <- tolower(names(stroke))
> head(stroke)
  sex       died       dstr age dgn coma diab minf han
1   1 7.01.1991 2.01.1991  76 INF    0    0    1   0
2   1      <NA> 3.01.1991  58 INF    0    0    0   0
3   1 2.06.1991 8.01.1991  74 INF    0    0    1   1
4   0 13.01.1991 11.01.1991 77 ICH   0    1    0   1
5   0 23.01.1996 13.01.1991 76 INF   0    1    0   1
6   1 13.01.1991 13.01.1991 48 ICH   1    0    0   1
```

(You can of course also just substitute the full path to stroke.csv instead
of using the system.file construction.)

In this data set, the two date variables died and dstr (date of stroke) appear as factor variables, which is the standard behaviour of read.table.
To convert them to class "Date", we use the function as.Date. This is
straightforward but requires some attention to the date format. The format used here is (day, month, year) separated by a period (dot character),
with year given as four digits. This is not a standard format, so we need
to specify it explicitly.

```
> stroke <- transform(stroke,
+    died = as.Date(died, format="%d.%m.%Y"),
+    dstr = as.Date(dstr, format="%d.%m.%Y"))
```

Notice the use of "percent-codes" to represent specific parts of the date:
%d indicates the day of the month, %m means the month as a number, and
%Y means that a four-digit year is used (notice the uppercase Y). The full
set of codes is documented on the help page for strptime.

Internally, dates are represented as the number of days before or after a
given point in time, known as the *epoch*. Specifically, the epoch is January
1, 1970, although this is an implementation detail that should not be relied
upon.

It is possible to perform arithmetic on dates; that is, they behave mostly
like numeric vectors:

```
> summary(stroke$died)
        Min.     1st Qu.      Median        Mean     3rd Qu.
"1991-01-07" "1992-03-14" "1993-01-23" "1993-02-15" "1993-11-04"
        Max.
"1996-02-22"
> summary(stroke$dstr)
        Min.     1st Qu.      Median        Mean     3rd Qu.
"1991-01-02" "1991-11-08" "1992-08-12" "1992-07-27" "1993-04-30"
        Max.
"1993-12-31"
```

```
> summary(stroke$died - stroke$dstr)
   Min. 1st Qu.  Median   Mean 3rd Qu.    Max.    NA's
    0.0     8.0    28.0   225.7   268.5  1836.0   338.0
> head(stroke$died - stroke$dstr)
Time differences in days
[1]    5    NA  145    2 1836    0
```

Notice that means and quantiles are displayed in date format (even if they are nonintegers). The count of NA values is not displayed for date variables even though the date of death is unknown for quite a few patients; this is a bit unfortunate, but it would conflict with a convention that numerical summaries have the same class as the object that is summarized (so you would get the count displayed as a date!).

The vector of differences between the two dates is actually an object of class "difftime". Such objects can have different units — when based on dates, it will always be "days", but for other kinds of time variables it can be "hours" or "seconds". Accordingly, it is somewhat bad practice just to treat the vector of differences as a numeric variable. The recommended procedure is to use as.numeric with an explicit units argument.

In the data file, NA for a death date means that the patient did not die before the end of the study on January 1, 1996. Six patients were recorded as having died after this date, but since there may well be unrecorded deaths among the remaining patients, we have to discard these death dates and just record the patients as alive at the end of the study.

We shall transform the data so that all patients have an end date plus an indicator of what happened at the end date: died or survived.

```
> stroke <- transform(stroke,
+   end = pmin(died,  as.Date("1996-1-1"), na.rm = T),
+   dead = !is.na(died) & died < as.Date("1996-1-1"))
> head(stroke)
  sex        died        dstr age dgn coma diab minf han
1   1 1991-01-07 1991-01-02  76 INF    0    0    1   0
2   1       <NA> 1991-01-03  58 INF    0    0    0   0
3   1 1991-06-02 1991-01-08  74 INF    0    0    1   1
4   0 1991-01-13 1991-01-11  77 ICH    0    1    0   1
5   0 1996-01-23 1991-01-13  76 INF    0    1    0   1
6   1 1991-01-13 1991-01-13  48 ICH    1    0    0   1
         end  dead
1 1991-01-07  TRUE
2 1996-01-01 FALSE
3 1991-06-02  TRUE
4 1991-01-13  TRUE
5 1996-01-01 FALSE
6 1991-01-13  TRUE
```

The pmin function calculates the minimum, but unlike the min function, which returns a single number, it does so in *parallel* across multiple vectors. The na.rm argument allows NA values to be ignored, so the result is that wherever died is missing or later than 1996-01-01, the end date becomes 1996-01-01 and the actual date of death otherwise.

The expression for dead is straightforward, although you should check that missing values are treated correctly. (They are. The & operator handles missingness such that if one argument is FALSE the result is FALSE, even if the other is NA.)

Finally, to obtain the observation time for all individuals, we can do

```
> stroke <- transform(stroke,
+    obstime = as.numeric(end - dstr, units="days")/365.25)
```

in which we pragmatically convert to "epidemiological years" of average length. (This cannot be done just by setting units="years". Objects of class "difftime" can only have units of "weeks" or less.)

Notice that we performed the transformations in three separate calls to transform. This was not just for the flow of the presentation; each of the last two calls refers to variables that were not defined previously. The transform function does not allow references to variables defined in the same call (we could have used within, though; see Section 2.1.8).

Further time classes

R also has classes that represent time to a granularity finer than 1 day. The "POSIXct" class (calendar time according to the POSIX standards) is similar to "Date" except that it counts seconds rather than days, and "POSIXlt" (local time) represents date and time using a structure that consists of fields for various components: year, month, day of month, hours, minutes, seconds, and more. Working with such objects involves, by and large, the same issues as for the "Date" class, although with a couple of extra twists related to time zones and Daylight Savings Time. We shall not go deeper into this area here.

10.1.4 Recoding multiple variables

In the previous sections, we had some cases where essentially the same transformation had to be applied to several variables. The solution in those cases was simply to repeat the operation, but it can happen that a data set contains many similar variables that all need to be recoded (questionnaire data may, for instance, have dozens of items rated on the same five-point scale). In such cases, you can make use of the fact that data frames are fundamentally lists and that `lapply` and indexing work on them. For instance, in dealing with the raw stroke data, we could have done the date handling as follows:

```
> rawstroke <- read.csv2(
+    system.file("rawdata","stroke.csv", package="ISwR"),
+    na.strings=".")
> ix <- c("DSTR", "DIED")
> rawstroke[ix] <- lapply(rawstroke[ix],
+                         as.Date, format="%d.%m.%Y")
> head(rawstroke)
  SEX        DIED        DSTR AGE DGN COMA DIAB MINF HAN
1   1 1991-01-07 1991-01-02  76 INF    0    0    1   0
2   1       <NA> 1991-01-03  58 INF    0    0    0   0
3   1 1991-06-02 1991-01-08  74 INF    0    0    1   1
4   0 1991-01-13 1991-01-11  77 ICH    0    1    0   1
5   0 1996-01-23 1991-01-13  76 INF    0    1    0   1
6   1 1991-01-13 1991-01-13  48 ICH    1    0    0   1
```

Similarly, the four binary variables could be converted to "No/Yes" factors in a single operation.

```
> ix <- 6:9
> rawstroke[ix] <- lapply(rawstroke[ix],
+                         factor, levels=0:1, labels=c("No","Yes"))
```

10.2   Conditional calculations

The `ifelse` function lets you apply different calculations to different parts of data. For illustration, we use a subset of the `stroke` data discussed in Section 10.1.3, but we use the "cooked" version contained in the `ISwR` package.

```
> strokesub <- ISwR::stroke[1:10,2:3]
> strokesub
         died        dstr
1  1991-01-07 1991-01-02
2        <NA> 1991-01-03
3  1991-06-02 1991-01-08
4  1991-01-13 1991-01-11
5        <NA> 1991-01-13
6  1991-01-13 1991-01-13
7  1993-12-01 1991-01-14
8  1991-12-12 1991-01-14
9        <NA> 1991-01-15
10 1993-11-10 1991-01-15
```

To compute the time on study and the event/censoring indicator needed for survival models, we can do as follows:

```
> strokesub <- transform(strokesub,
+    event = !is.na(died))
> strokesub <- transform(strokesub,
+   obstime = ifelse(event, died-dstr, as.Date("1996-1-1") - dstr))
> strokesub
          died        dstr event obstime
1   1991-01-07 1991-01-02  TRUE       5
2         <NA> 1991-01-03 FALSE    1824
3   1991-06-02 1991-01-08  TRUE     145
4   1991-01-13 1991-01-11  TRUE       2
5         <NA> 1991-01-13 FALSE    1814
6   1991-01-13 1991-01-13  TRUE       0
7   1993-12-01 1991-01-14  TRUE    1052
8   1991-12-12 1991-01-14  TRUE     332
9         <NA> 1991-01-15 FALSE    1812
10  1993-11-10 1991-01-15  TRUE    1030
```

The way ifelse works is that it takes three arguments: test, yes, and no. All three are vectors of the same length (if not, they will be made so by recycling). The answer is "stitched together" of pieces of yes and no in the sense that the yes element is selected wherever test is TRUE and the no element where it is FALSE. When the condition is NA, so is the result.

Notice that both alternatives are computed (exceptions are made for the cases where the condition is all TRUE or all FALSE). This is not usually a problem in terms of speed, but it does mean that ifelse is not the right tool to use if you want to avoid, for example, taking the logarithm of negative values. Also notice that ifelse discards attributes, including the class, so that obstime is not of class "difftime" even though both the yes and the no part are. This sometimes makes using ifelse more trouble than it is worth, and it can be preferable simply to use explicit subsetting operations.

10.3   Combining and restructuring data frames

In this section, we discuss ways of joining data frames either "vertically" (adding records) or "horizontally" (adding variables). We also look at the issue of converting data with repeated measurements of the same variables between the "long" and the "wide" formats.

10.3.1   Appending frames

Sometimes data are received from multiple sources and you need to combine them to form one bigger data set. In this subsection, we consider the case where data are combined by "vertical stacking"; that is, you start out with data frames which refer to separate rows of the result — typically different subjects. It is required that the data frames contain the same variables, although not necessarily in the same order (this is unlike some other statistical systems, which will simply insert missing values for variables that are absent in a data set).

To simulate such a situation, suppose that the `juul` data set had been collected separately for boys and girls. In that case, the data frames might not contain the variable `sex`, since this is the same for everyone in the same data frame, and variables that only make sense for one gender may also have been omitted for the other group.

```
> juulgrl <- subset(juul, sex==2, select=-c(testvol,sex))
> juulboy <- subset(juul, sex==1, select=-c(menarche,sex))
```

Notice the use of the `select` argument to `subset`. The processing of this argument replaces column names by column numbers, and the resulting expression is used to index the data frame. The net effect of the negative indices is to remove, for example, `testvol` and `sex` from `juulgrl`.

To put the data frames back together, you must first add in the missing variables

```
> juulgrl$sex <- factor("F")
> juulgrl$testvol <- NA
> juulboy$sex <- factor("M")
> juulboy$menarche <- NA
```

and then it is just a matter of using the `rbind` method for data frames:

```
> juulall <- rbind(juulboy, juulgrl)
> names(juulall)
[1] "age"       "igf1"       "tanner"     "testvol"    "sex"
[6] "menarche"
```

Notice that `rbind` uses the column names (so that it does not concatenate unrelated variables even though the order of columns differs in the two data frames) and that the order of variables in the first data frame "wins": The result has the variables in the same order as `juulboy`. Notice also that `rbind` is being smart about factor levels:

```
> levels(juulall$sex)
[1] "M" "F"
```

10.3.2 Merging data frames

Just as you may have different groups of subjects collected in separate data sets, you may also have different sorts of data on the same patients collected separately. For example, you could have one data frame with registry data, one with clinical biochemistry data, and one with questionnaire data. It may work to use `cbind` to stick the data frames together side-by-side, but it could be dangerous: What if the data are not complete in all data frames or out of sequence? You typically have to work with a unique subject identification code to avoid mistakes of this sort.

The `merge` function deals with these issues. It works by matching on one or several variables from each data frame. By default, this is the set of variables that have the same name in both frames (typically, there is a variable called something like `ID`, which holds the subject identification). Assuming that this default works and that the two data frames are called respectively `dfx` and `dfy`, the merged frame is computed simply as

```
merge(dfx, dfy)
```

However, there may be variables of the same name in both frames. In such cases, you can add a `by` argument, which contains the variable name or names to match on as in

```
merge(dfx, dfy, by="ID")
```

Any other variables that appear in both frames will have `.x` or `.y` appended to their name in the result. It is recommended to use this format in any case as a safeguard and for readability and explicitness. If the matching variable(s) have different names in the two data frames, you can use `by.x` and `by.y`.

Matching is not necessarily one-to-one. One of the data sets might for instance hold tabular material corresponding to the study population. The common example is mortality tables. In such cases, there is generally a many-to-one relationship between the data frames. More than one subject in the study population will belong to the table entry for 40–49 year-olds, and the rows of the table will have to be duplicated accordingly during the merge.

To illustrate these concepts, we use the data set `nickel`. This describes a cohort of nickel smelting workers in South Wales. The data set `ewrates` contains a table of the population mortality by year and age group in five-year intervals.

```
> head(nickel)
  id icd exposure      dob  age1st    agein   ageout
1  3   0        5 1889.019 17.4808  45.2273  92.9808
```

```
2   4 162         5 1885.978 23.1864 48.2684 63.2712
3   6 163        10 1881.255 25.2452 52.9917 54.1644
4   8 527         9 1886.340 24.7206 47.9067 69.6794
5   9 150         0 1879.500 29.9575 54.7465 76.8442
6  10 163         2 1889.915 21.2877 44.3314 62.5413
> head(ewrates)
  year age lung nasal other
1 1931  10    1     0  1269
2 1931  15    2     0  2201
3 1931  20    6     0  3116
4 1931  25   14     0  3024
5 1931  30   30     1  3188
6 1931  35   68     1  4165
```

Suppose we wish to merge these two data sets according to the values at entry into the study population. This age is contained in agein, and the date of entry is computed as dob + agein. You can compute group codes corresponding to ewrates as follows:

```
> nickel <- transform(nickel,
+     agr = trunc(agein/5)*5,
+     ygr = trunc((dob+agein-1)/5)*5+1)
```

The trunc function rounds values towards zero. Notice that the age groups start on values that are evenly divisible by 5, whereas the year groups end on such values; this is why the expression for ygr subtracts 1 and adds it back after truncation. (Actually this does not matter because all enrollment dates were April 1 of 1934, 1939, 1944, or 1949.) Notice also that we do not use the same variable names as in ewrates. We could have done so, but the names age and year would be unintuitive in the context of the nickel data.

With the age and year groups defined, it is an easy matter to perform the merge. We just need to account for the fact that we have used different variable names in the two data frames.

```
> mrg <- merge(nickel, ewrates,
+     by.x=c("agr","ygr"), by.y=c("age","year"))
> head(mrg,10)
   agr  ygr  id icd exposure      dob   age1st    agein   ageout
1   20 1931 273 154        0 1909.500 14.6913  24.7465  55.9302
2   20 1931 213 162        0 1910.129 14.2018  24.1177  63.0493
3   20 1931 546   0        0 1909.500 14.4945  24.7465  72.5000
4   20 1931 574 491        0 1909.729 14.0356  24.5177  70.6592
5   20 1931 110   0        0 1909.247 14.0302  24.9999  72.7534
6   20 1931 325 434        0 1910.500 14.0737  23.7465  43.0343
7   25 1931  56 502        2 1904.500 18.2917  29.7465  51.5847
8   25 1931 690 420        0 1906.500 17.2206  27.7465  55.1219
9   25 1931 443 420        0 1905.326 14.5562  28.9204  65.7616
10  25 1931 137 465        0 1905.386 19.0808  28.8601  74.2794
   lung nasal other
```

```
1      6     0   3116
2      6     0   3116
3      6     0   3116
4      6     0   3116
5      6     0   3116
6      6     0   3116
7     14     0   3024
8     14     0   3024
9     14     0   3024
10    14     0   3024
```

We have only described the main function of merge. There are also options to include rows that only exist in one of the two frames (all, all.x,
all.y), and it may also be useful to know that the pseudo-variable
row.names will allow matching on row names.

We have discussed the cases of one-to-one and many-to-one matching.
Many-to-many is possible but rarely useful. What happens in that case
is that the "Cartesian product" is formed by generating all combinations
of rows from the two frames within each matching set. The extreme case
of many-to-many matching occurs if the by set is empty, which gives a
result with as many rows as the *product* of the row counts. This sometimes
surprises people who expect that the row number will act as an implicit
ID.

10.3.3 Reshaping data frames

Longitudinal data come in two different forms: a "wide" format, where
there is a separate column for each time point but only one record per
case; and a "long" format, where there are multiple records for each case,
one for each time point. The long format is more general since it does not
need to assume that the cases are recorded at the same set of times, but
when applicable it may be easier to work with data in the wide format,
and some statistical functions expect it that way. Other functions expect to
find data in the long format. Either way, there is a need to convert from
one format to another. This is what the reshape function does.

Consider the following data from a randomized study of bone metabolism
data during Tamoxifen treatment after breast cancer. The concentration
of alkaline phosphatase is recorded at baseline and 3, 6, 9, 12, 18, and
24 months after treatment start.

```
> head(alkfos)
  grp  c0  c3  c6  c9 c12 c18 c24
1   1 142 140 159 162 152 175 148
2   1 120 126 120 146 134 119 116
3   1 175 161 168 164 213 194 221
```

```
4    1 234 203 174 197 289 174 189
5    1  94 107 146 124 128  98 114
6    1 128  97 113 203  NA  NA  NA
```

In the simplest uses of reshape, the function will assume that the variable names encode the information necessary for reshaping to the long format. By default, it assumes that variable names are separated from time of measurement by a " . " (dot), so we might oblige by modifying the name format.

```
> a2 <- alkfos
> names(a2) <- sub("c", "c.", names(a2))
> names(a2)
[1] "grp"  "c.0"  "c.3"  "c.6"  "c.9"  "c.12" "c.18" "c.24"
```

The sub function does substitutions within character strings, in this case replacing the string "c" with "c.". Alternatively, the original name format (c0, ..., c24) can be handled by adding sep="" to the reshape call.

Once we have the variable naming in place, the only things we need to specify are the direction of the reshape and the set of variables to be considered time-varying. As a convenience feature, the latter can be specified by index rather than by name.

```
> a.long <- reshape(a2, varying=2:8, direction="long")
> head(a.long)
     grp time   c id
1.0   1    0 142  1
2.0   1    0 120  2
3.0   1    0 175  3
4.0   1    0 234  4
5.0   1    0  94  5
6.0   1    0 128  6
> tail(a.long)
      grp time   c id
38.24   2   24  95 38
39.24   2   24  NA 39
40.24   2   24 192 40
41.24   2   24  94 41
42.24   2   24 194 42
43.24   2   24 129 43
```

Notice that the sort order of the result is that id varies within time. This is the most convenient format to generate technically, but if you prefer the opposite sort order, just use

```
> o <- with(a.long, order(id, time))
> head(a.long[o,], 10)
     grp time   c id
```

```
1.0     1     0 142   1
1.3     1     3 140   1
1.6     1     6 159   1
1.9     1     9 162   1
1.12    1    12 152   1
1.18    1    18 175   1
1.24    1    24 148   1
2.0     1     0 120   2
2.3     1     3 126   2
2.6     1     6 120   2
```

To demonstrate the reverse procedure, we use the same data, in the long format. Actually, this is a bit too easy because reshape has inserted enough information in its output to let you convert to the wide format just by saying reshape(a.long). To simulate the situation where the original data are given in the long format, we remove the "reshapeLong" attribute, which holds these data. Furthermore, we remove the records for which we have missing data by using na.omit.

```
> a.long2 <- na.omit(a.long)
> attr(a.long2, "reshapeLong") <- NULL
```

To convert a.long2 to the wide format, use

```
> a.wide2 <- reshape(a.long2, direction="wide", v.names="c",
+                    idvar="id", timevar="time")
> head(a.wide2)
    grp id c.0 c.3 c.6 c.9 c.12 c.18 c.24
1.0   1  1 142 140 159 162  152  175  148
2.0   1  2 120 126 120 146  134  119  116
3.0   1  3 175 161 168 164  213  194  221
4.0   1  4 234 203 174 197  289  174  189
5.0   1  5  94 107 146 124  128   98  114
6.0   1  6 128  97 113 203   NA   NA   NA
```

Notice that NA values are filled in for patient no. 6, for whom only the first four observations are available.

The arguments idvar and timevar specify the names of the variables that contain the ID and the time for each observation. It is not strictly necessary to specify them if they have their default names, but it is good practice to do so. The argument v.names specifies the time-varying variables; notice that if it were omitted, then the grp variable would also be treated as time-varying.

10.4   Per-group and per-case procedures

A specific data management task involves operations within subsets of a data frame, particularly those where there are multiple records for each individual. Examples include calculation of cumulative dosage in a pharmacokinetic experiment and various methods of normalization and standardization.

A nice general approach to such tasks is first to split the data into a list of groups, operate on each group, and then put the pieces back together.

Consider the task of normalizing the values of alkaline phosphatase in a.long to their baseline values. The split function can be used to generate a list of the individual time courses:

```
> l <- split(a.long$c, a.long$id)
> l[1:3]
$`1`
[1] 142 140 159 162 152 175 148

$`2`
[1] 120 126 120 146 134 119 116

$`3`
[1] 175 161 168 164 213 194 221
```

Next, we apply a function to each element of the list and collect the results using lapply.

```
> l2 <- lapply(l, function(x) x / x[1])
```

Finally, we put the pieces back together using unsplit, which is the reverse operation of split. Notice that a.long has id varying within time, so this is not just a matter of concatenating the elements of l2. The data for the first patient are now

```
> a.long$c.adj <- unsplit(l2, a.long$id)
> subset(a.long, id==1)
      grp time   c id      c.adj
1.0     1    0 142  1 1.0000000
1.3     1    3 140  1 0.9859155
1.6     1    6 159  1 1.1197183
1.9     1    9 162  1 1.1408451
1.12    1   12 152  1 1.0704225
1.18    1   18 175  1 1.2323944
1.24    1   24 148  1 1.0422535
```

In fact, there is a function that formalizes this sort of split-modify-unsplit operation. It is called ave because the default use is to replace data with group averages, but it can also be used for more general transformations. The following is an alternative way of doing the same computation as above:

```
> a.long$c.adj <- ave(a.long$c, a.long$id,
+      FUN = function(x) x / x[1])
```

In the preceding code, we worked on the single vector a.long$c. Alternatively, we can split the entire data frame and use code like

```
> l <- split(a.long, a.long$id)
> l2 <- lapply(l, transform, c.adj = c / c[1])
> a.long2 <- unsplit(l2, a.long$id)
```

Notice how the last argument to lapply is passed on to transform, so that you effectively call transform(x, c.adj = c / c[1]) for each data frame x in the list l. This procedure is somewhat less efficient than the first one because there is more copying of data, but it generalizes to more complex transformations.

10.5   Time splitting

This section is rather advanced, and the beginner may want to skip it on the first read. Understanding the contents is not crucial for the later parts of the book. On the other hand, apart from solving the particular problem, this is also a rather nice first example of the use of ad hoc programming in R and also of the "lateral thinking" that is sometimes required.

The merge operation of the nickel and ewrates data in Section 10.3.2 does not really make sense statistically: We merged in the mortality table corresponding to the age at the time of entry into the study population. However, the data set is about cancer, a slow disease, and an exposure that perhaps leads to an increased risk 20 or more years later. If the subjects typically die around age 50, the population mortality for people of age 30 is hardly relevant.

A sensible statistical analysis needs to consider the population mortality during the entire follow-up period. One way to handle this issue is to split the individuals into multiple "sub-individuals".

In the data set, the first six observations are (after the merge in Section 10.3.2)

```
> head(nickel)
  id icd exposure       dob  age1st    agein  ageout agr  ygr
1  3   0        5  1889.019 17.4808 45.2273 92.9808  45 1931
```

```
2   4 162          5 1885.978 23.1864 48.2684 63.2712   45 1931
3   6 163         10 1881.255 25.2452 52.9917 54.1644   50 1931
4   8 527          9 1886.340 24.7206 47.9067 69.6794   45 1931
5   9 150          0 1879.500 29.9575 54.7465 76.8442   50 1931
6  10 163          2 1889.915 21.2877 44.3314 62.5413   40 1931
```

Consider the individual with `id == 4`; this person entered the study at the age of 48.2684 and died (from lung cancer) at the age of 63.2712 (apologies for the excess precision). The time-splitting method treats this subject as four separate subjects, one entering the study at age 48.2684 and leaving at age 50 (on his 50th birthday) and the others covering the intervals 50–55, 55–60, and 60–63.2712. The first three are censored observations, as the subject did not die.

If we merge these data with the population tables, then we can compute the expected number of deaths in a given age interval and compare that with the actual number of deaths.

Taking advantage of the vectorized nature of computations in R, the nice way of doing this is to loop over age intervals, "trimming" every observation period to each interval.

To trim the observation periods to ages between (say) 60 and 65, the entry and exit times should be adjusted to the interval if they fall outside of it, cases that are unobserved during the interval should be removed, and if the subject did not die inside the interval, `icd` should be set to 0.

The easiest procedure is to "shoot first and ask later". The adjusted entry and exit times are

```
> entry <- pmax(nickel$agein, 60)
> exit <- pmin(nickel$ageout, 65)
```

or rather they would be if there were always a suitable overlap between the observation period and the target age interval. However, there are people leaving the study population before age 60 (by death or otherwise) and people entering the study after age 65. In either case, what goes wrong is that `entry >= exit`, and we can check for such cases by calculating

```
> valid <- (entry < exit)
> entry <- entry[valid]
> exit  <- exit[valid]
```

The censoring indicator for valid cases is

```
> cens <- (nickel$ageout[valid] > 65)
```

(We might have used `cens <- (exit == 65)`, but it is a good rule to avoid testing floating point data for equality.)

The trimmed data set can then be obtained as

```
> nickel60 <- nickel[valid,]
> nickel60$icd[cens] <- 0
> nickel60$agein <- entry
> nickel60$ageout <- exit
> nickel60$agr <- 60
> nickel60$ygr <- with(nickel60, trunc((dob+agein-1)/5)*5+1)
```

and the first lines of the result are

```
> head(nickel60)
  id icd exposure      dob  agelst agein  ageout agr  ygr
1  3   0        5 1889.019 17.4808    60 65.0000  60 1946
2  4 162        5 1885.978 23.1864    60 63.2712  60 1941
4  8   0        9 1886.340 24.7206    60 65.0000  60 1946
5  9   0        0 1879.500 29.9575    60 65.0000  60 1936
6 10 163        2 1889.915 21.2877    60 62.5413  60 1946
7 15 334        0 1890.500 23.2836    60 62.0000  60 1946
```

A couple of fine points: If someone dies exactly at age 65, they are counted as dying inside the age interval. Conversely, we do not include people dying exactly at age 60; they belong in the interval 55–60 (for purposes like those of Chapter 15, one should avoid observation intervals of length zero). It was also necessary to recompute `ygr` since this was based on the original `agein`.

To get the fully expanded data set, you could repeat the above for each age interval (20–25, …, 95–100) and append the resulting 16 data frames with `rbind`. However, this gets rather long-winded, and there is a substantial risk of copy-paste errors. Instead, you can do a little programming. First, wrap up the procedure for one group as a function:

```
> trim <- function(start)
+ {
+    end   <- start + 5
+    entry <- pmax(nickel$agein, start)
+    exit  <- pmin(nickel$ageout, end)
+    valid <- (entry < exit)
+    cens  <- (nickel$ageout[valid] > end)
+    result <- nickel[valid,]
+    result$icd[cens] <- 0
+    result$agein <- entry[valid]
+    result$ageout <- exit[valid]
+    result$agr <- start
+    result$ygr <- with(result, trunc((dob+agein-1)/5)*5+1)
+    result
+ }
```

(In practice, you should not type all this at the command line but use a script window or an editor; see Section 2.1.3.)

This is typical ad hoc programming. The function is far from general since it relies on knowing various names, and it also hardcodes the interval length as 5. However, more generality is not required for a one-off calculation. The important thing for the purpose at hand is to make the dependence on `start` explicit so that we can loop over it.

With this definition, `trim(60)` is equivalent to the `nickel60` we computed earlier:

```
> head(trim(60))
  id icd exposure      dob  age1st agein  ageout agr  ygr
1  3   0          5 1889.019 17.4808    60 65.0000  60 1946
2  4 162          5 1885.978 23.1864    60 63.2712  60 1941
4  8   0          9 1886.340 24.7206    60 65.0000  60 1946
5  9   0          0 1879.500 29.9575    60 65.0000  60 1936
6 10 163          2 1889.915 21.2877    60 62.5413  60 1946
7 15 334          0 1890.500 23.2836    60 62.0000  60 1946
```

To get results for all intervals, do the following:

```
> nickel.expand <- do.call("rbind", lapply(seq(20,95,5), trim))
> head(nickel.expand)
     id icd exposure      dob  age1st   agein ageout agr  ygr
84  110   0          0 1909.247 14.0302 24.9999     25  20 1931
156 213   0          0 1910.129 14.2018 24.1177     25  20 1931
197 273   0          0 1909.500 14.6913 24.7465     25  20 1931
236 325   0          0 1910.500 14.0737 23.7465     25  20 1931
384 546   0          0 1909.500 14.4945 24.7465     25  20 1931
400 574   0          0 1909.729 14.0356 24.5177     25  20 1931
```

The `do.call` construct works by creating a call to `rbind` with a given argument list, which in this case is the return value from `lapply`, which in turn has applied the `trim` function to each of the values 20, 25, ... 95. That is, the whole thing is equivalent to

```
rbind(trim(20), trim(25), ......, trim(95))
```

Displaying the result for a single subject yields, for example,

```
> subset(nickel.expand, id==4)
      id icd exposure      dob  age1st   agein  ageout agr  ygr
2      4   0          5 1885.978 23.1864 48.2684 50.0000  45 1931
2100   4   0          5 1885.978 23.1864 50.0000 55.0000  50 1931
2102   4   0          5 1885.978 23.1864 55.0000 60.0000  55 1936
2104   4 162          5 1885.978 23.1864 60.0000 63.2712  60 1941
```

(The strange row names occur because multiple data frames with the same row names are being `rbind`-ed together and data frames must have unique row names.)

A weakness of the `ygr` computation is that since `ygr` refers to the calendar time group at `agein`, it may be off by up to 5 years. However, lung cancer death rates by age do not change that quickly, so we leave it at this. A more careful procedure, and in fact the common practice in epidemiology, is to split on both age and calendar time. The `Epi` package contains generalized time-splitters `splitLexis` and `cutLexis`, which are useful for this purpose and also for handling the related case of splitting time based on individual events (e.g., childbirth).

As a final step, we can merge in the mortality table as we did in Section 10.3.2.

```
> nickel.expand <- merge(nickel.expand, ewrates,
+    by.x=c("agr","ygr"), by.y=c("age","year"))
> head(nickel.expand)
  agr  ygr  id icd exposure    dob    agelst    agein ageout lung
1  20 1931 325   0        0 1910.500 14.0737 23.7465     25    6
2  20 1931 273   0        0 1909.500 14.6913 24.7465     25    6
3  20 1931 110   0        0 1909.247 14.0302 24.9999     25    6
4  20 1931 574   0        0 1909.729 14.0356 24.5177     25    6
5  20 1931 213   0        0 1910.129 14.2018 24.1177     25    6
6  20 1931 546   0        0 1909.500 14.4945 24.7465     25    6
  nasal other
1     0  3116
2     0  3116
3     0  3116
4     0  3116
5     0  3116
6     0  3116
```

For later use, the expanded data set is made available "precooked" in the `ISwR` package under the name `nickel.expand`. We return to the data set in connection with the analysis of rates in Chapter 15.

10.6 Exercises

10.1 Create a factor in which the `blood.glucose` variable in the `thuesen` data is divided into the intervals $(4,7], (7,9], (9,12],$ and $(12,20]$. Change the level names to "low", "intermediate", "high", and "very high".

10.2 In the `bcmort` data set, the four-level factor `cohort` can be considered the product of two two-level factors, say `period` and `area`. How can you generate them?

10.3 Convert the `ashina` data to the long format. Consider how to encode whether the `vas` measurement is from the first or the second measurement session.

10.4 Split the `stroke` data according to `obsmonths` into time intervals 0–0.5, 0.5–2, 2–12, and 12+ months after stroke.

11

Multiple regression

This chapter discusses the case of regression analysis with multiple predictors. There is not really much new here since model specification and output do not differ a lot from what has been described for regression analysis and analysis of variance. The news is mainly the model search aspect, namely among a set of potential descriptive variables to look for a subset that describes the response sufficiently well.

The basic model for multiple regression analysis is

$$y = \beta_0 + \beta_1 x_1 + \cdots + \beta_k x_k + \epsilon$$

where $x_1, \ldots x_k$ are explanatory variables (also called predictors) and the parameters $\beta_1, \ldots, \beta_k$ can be estimated using the method of least squares (see Section 6.1). A closed-form expression for the estimates can be derived using matrix calculus, but we do not go into the details of that here.

11.1 Plotting multivariate data

As an example in this chapter, we use a study concerning lung function in patients with cystic fibrosis in Altman (1991, p. 338). The data are in the cystfibr data frame in the ISwR package.

P. Dalgaard, *Introductory Statistics with R*,
DOI: 10.1007/978-0-387-79054-1_11, © Springer Science+Business Media, LLC 2008

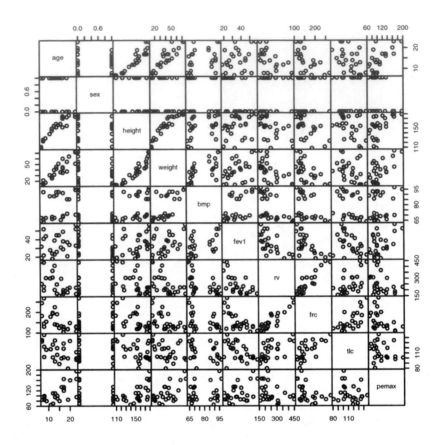

Figure 11.1. Pairwise plots for cystic fibrosis data.

You can obtain pairwise scatterplots between all the variables in the data set. This is done using the function `pairs`. To get Figure 11.1, you simply write

```
> par(mex=0.5)
> pairs(cystfibr, gap=0, cex.labels=0.9)
```

The arguments `gap` and `cex.labels` control the visual appearance by removing the space between subplots and decreasing the font size. The `mex` graphics parameter reduces the interline distance in the margins.

A similar plot is obtained by simply saying `plot(cystfibr)` since the `plot` function is generic and behaves differently depending on the class of its arguments (see Section 2.3.2). Here the argument is a data frame and a `pairs` plot is a fairly reasonable thing to get when asking for a plot of an entire data frame (although you might equally reasonably have expected a histogram or a barchart of each variable instead).

The individual plots do get rather small, probably not suitable for direct publication, but such plots are quite an effective way of obtaining an overview of multidimensional issues. For example, the close relations among age, height, and weight appear clearly on the plot.

In order to be able to refer directly to the variables in cystfibr, we add it to the search path (a harmless warning about masking of tlc ensues at this point):

```
> attach(cystfibr)
```

Because this data set contains common variable names such as age, height, and weight, it is a good idea to ensure that you do not have identically named variables in the workspace at this point. In particular, such names were used in the introductory session.

11.2   Model specification and output

Specification of a multiple regression analysis is done by setting up a model formula with + between the explanatory variables:

```
lm(pemax~age+sex+height+weight+bmp+fev1+rv+frc+tlc)
```

which is meant to be read as "pemax is described using a model that is additive in age, sex, and so forth." (pemax is the maximal expiratory pressure. See Appendix B for a description of the other variables in cystfibr.)

As usual, there is not much output from lm itself, but with the aid of summary you can obtain some more interesting output:

```
> summary(lm(pemax~age+sex+height+weight+bmp+fev1+rv+frc+tlc))

Call:
lm(formula = pemax ~ age + sex + height + weight + bmp + fev1 +
    rv + frc + tlc)

Residuals:
    Min      1Q  Median      3Q     Max
-37.338 -11.532   1.081  13.386  33.405

Coefficients:
            Estimate Std. Error t value Pr(>|t|)
(Intercept) 176.0582   225.8912   0.779    0.448
```

```
age             -2.5420     4.8017  -0.529    0.604
sex             -3.7368    15.4598  -0.242    0.812
height          -0.4463     0.9034  -0.494    0.628
weight           2.9928     2.0080   1.490    0.157
bmp             -1.7449     1.1552  -1.510    0.152
fev1             1.0807     1.0809   1.000    0.333
rv               0.1970     0.1962   1.004    0.331
frc             -0.3084     0.4924  -0.626    0.540
tlc              0.1886     0.4997   0.377    0.711
```

```
Residual standard error: 25.47 on 15 degrees of freedom
Multiple R-squared: 0.6373,     Adjusted R-squared: 0.4197
F-statistic: 2.929 on 9 and 15 DF, p-value: 0.03195
```

The layout should be well known by now. Notice that there is not one single significant t value, but the joint F test is nevertheless significant, so there must be an effect somewhere. The reason is that the t tests only say something about what happens if you remove one variable and leave in all the others. You cannot see whether a variable would be statistically significant in a reduced model; all you can see is that no variable *must* be included.

Note further that there is quite a large difference between the unadjusted and the adjusted R^2, which is due to the large number of variables relative to the number of degrees of freedom for the variance. Recall that the former is the change in residual sum of squares relative to an empty model, whereas the latter is the similar change in residual *variance*:

```
> 1-25.5^2/var(pemax)
[1] 0.4183949
```

The 25.5 comes from "residual standard error" in the `summary` output.

The ANOVA table for a multiple regression analysis is obtained using `anova` and gives a rather different picture:

```
> anova(lm(pemax~age+sex+height+weight+bmp+fev1+rv+frc+tlc))
Analysis of Variance Table
```

```
Response: pemax
          Df  Sum Sq Mean Sq F value   Pr(>F)
age        1 10098.5 10098.5 15.5661 0.001296 **
sex        1   955.4   955.4  1.4727 0.243680
height     1   155.0   155.0  0.2389 0.632089
weight     1   632.3   632.3  0.9747 0.339170
bmp        1  2862.2  2862.2  4.4119 0.053010 .
fev1       1  1549.1  1549.1  2.3878 0.143120
rv         1   561.9   561.9  0.8662 0.366757
frc        1   194.6   194.6  0.2999 0.592007
tlc        1    92.4    92.4  0.1424 0.711160
Residuals 15  9731.2   648.7
```

```
---
Signif. codes:   0 '***' 0.001 '**' 0.01 '*' 0.05 '.' 0.1 ' ' 1
```

Note that, except for the very last line ("tlc"), there is practically no correspondence between these F tests and the t tests from summary. In particular, the effect of age is now significant. That is because these tests are successive; they correspond to (reading upward from the bottom) a stepwise removal of terms from the model until finally only age is left. During the process, bmp came close to the magical 5% limit, but in view of the number of tests, this is hardly noteworthy.

The probability that one out of eight independent tests gives a p-value of 0.053 or below is actually just over 35%! The tests in the ANOVA table are not completely independent, but the approximation should be good.

The ANOVA table indicates that there is no significant improvement of the model once age is included. It is possible to perform a joint test for whether *all* the other variables can be removed by adding up the sums of squares contributions and using the sum for an F test; that is,

```
> 955.4+155.0+632.3+2862.2+1549.1+561.9+194.6+92.4
[1] 7002.9
> 7002.9/8
[1] 875.3625
> 875.36/648.7
[1] 1.349407
> 1-pf(1.349407,8,15)
[1] 0.2935148
```

This corresponds to collapsing the eight lines of the table so that it would look like this:

```
            Df   Sum Sq   Mean Sq       F   Pr(>F)
age          1  10098.5   10098.5  15.566  0.00130
others       8   7002.9     875.4   1.349  0.29351
Residual    15   9731.2     648.7
```

(Note that this is "cheat output", in which we have manually inserted the numbers computed above.)

A procedure leading directly to the result is

```
> m1<-lm(pemax~age+sex+height+weight+bmp+fev1+rv+frc+tlc)
> m2<-lm(pemax~age)
> anova(m1,m2)
Analysis of Variance Table

Model 1: pemax ~ age + sex + height + weight + bmp + fev1 + rv +
    frc + tlc
Model 2: pemax ~ age
```

```
   Res.Df      RSS Df Sum of Sq       F Pr(>F)
1     15   9731.2
2     23 16734.2 -8   -7002.9 1.3493 0.2936
```

which gives the appropriate F test with no manual computation.

Notice, however, that you need to be careful to ensure that the two models are actually nested. R does not check this, although it does verify that the number of response observations is the same to safeguard against the more obvious mistakes. (When there are missing values in the descriptive variables, it's easy for the smaller model to contain more data points.)

From the ANOVA table, we can thus see that it is allowable to remove all variables except `age`. However, that this particular variable is left in the model is primarily due to the fact that it was mentioned first in the model specification, as we see below.

11.3   Model search

R has the `step()` function for performing model searches by the Akaike information criterion. Since that is well beyond the scope of this book, we use simple manual variants of backwards elimination.

In the following, we go through a practical model reduction for the example data. Notice that the output has been slightly edited to take up less space.

```
> summary(lm(pemax~age+sex+height+weight+bmp+fev1+rv+frc+tlc))
...
            Estimate Std. Error t value Pr(>|t|)
(Intercept) 176.0582   225.8912   0.779    0.448
age          -2.5420     4.8017  -0.529    0.604
sex          -3.7368    15.4598  -0.242    0.812
height       -0.4463     0.9034  -0.494    0.628
weight        2.9928     2.0080   1.490    0.157
bmp          -1.7449     1.1552  -1.510    0.152
fev1          1.0807     1.0809   1.000    0.333
rv            0.1970     0.1962   1.004    0.331
frc          -0.3084     0.4924  -0.626    0.540
tlc           0.1886     0.4997   0.377    0.711
...
```

One advantage of doing model reductions by hand is that you may impose some logical structure on the process. In the present case, it may, for instance, be natural to try to remove other lung function indicators first.

```
> summary(lm(pemax~age+sex+height+weight+bmp+fev1+rv+frc))
```

```
...
              Estimate Std. Error t value Pr(>|t|)
(Intercept)  221.8055    185.4350   1.196   0.2491
age           -3.1346      4.4144  -0.710   0.4879
sex           -4.6933     14.8363  -0.316   0.7558
height        -0.5428      0.8428  -0.644   0.5286
weight         3.3157      1.7672   1.876   0.0790 .
bmp           -1.9403      1.0047  -1.931   0.0714 .
fev1           1.0183      1.0392   0.980   0.3417
rv             0.1857      0.1887   0.984   0.3396
frc           -0.2605      0.4628  -0.563   0.5813
...
> summary(lm(pemax~age+sex+height+weight+bmp+fev1+rv))
...
              Estimate Std. Error t value Pr(>|t|)
(Intercept)  166.71822  154.31294   1.080   0.2951
age           -1.81783    3.66773  -0.496   0.6265
sex            0.10239   11.89990   0.009   0.9932
height        -0.40981    0.79257  -0.517   0.6118
weight         2.87386    1.55120   1.853   0.0814 .
bmp           -1.94971    0.98415  -1.981   0.0640 .
fev1           1.41526    0.74788   1.892   0.0756 .
rv             0.09567    0.09798   0.976   0.3425
...
> summary(lm(pemax~age+sex+height+weight+bmp+fev1))
...
              Estimate Std. Error t value Pr(>|t|)
(Intercept)  260.6313   120.5215   2.163   0.0443 *
age           -2.9062      3.4898  -0.833   0.4159
sex           -1.2115     11.8083  -0.103   0.9194
height        -0.6067      0.7655  -0.793   0.4384
weight         3.3463      1.4719   2.273   0.0355 *
bmp           -2.3042      0.9136  -2.522   0.0213 *
fev1           1.0274      0.6329   1.623   0.1219
...
> summary(lm(pemax~age+sex+height+weight+bmp))
...
              Estimate Std. Error t value Pr(>|t|)
(Intercept)  280.4482   124.9556   2.244   0.0369 *
age           -3.0750      3.6352  -0.846   0.4081
sex          -11.5281     10.3720  -1.111   0.2802
height        -0.6853      0.7962  -0.861   0.4001
weight         3.5546      1.5281   2.326   0.0312 *
bmp           -1.9613      0.9263  -2.117   0.0476 *
...
```

As is seen, there was no obstacle to removing the four lung function variables. Next we try to reduce among the variables that describe the patient's state of physical development or size. Initially, we avoid removing `weight` and `bmp` since they appear to be close to the 5% significance limit.

```
> summary(lm(pemax~age+height+weight+bmp))
...
              Estimate Std. Error t value Pr(>|t|)
(Intercept)  274.5307   125.5745    2.186   0.0409 *
age           -3.0832     3.6566   -0.843   0.4091
height        -0.6985     0.8008   -0.872   0.3934
weight         3.6338     1.5354    2.367   0.0282 *
bmp           -1.9621     0.9317   -2.106   0.0480 *
...
> summary(lm(pemax~height+weight+bmp))
...
              Estimate Std. Error t value Pr(>|t|)
(Intercept)  245.3936   119.8927    2.047   0.0534 .
height        -0.8264     0.7808   -1.058   0.3019
weight         2.7717     1.1377    2.436   0.0238 *
bmp           -1.4876     0.7375   -2.017   0.0566 .
...
> summary(lm(pemax~weight+bmp))
...
              Estimate Std. Error t value Pr(>|t|)
(Intercept)  124.8297    37.4786    3.331 0.003033 **
weight         1.6403     0.3900    4.206 0.000365 ***
bmp           -1.0054     0.5814   -1.729 0.097797 .
...
> summary(lm(pemax~weight))
...
              Estimate Std. Error t value Pr(>|t|)
(Intercept)   63.5456    12.7016    5.003 4.63e-05 ***
weight         1.1867     0.3009    3.944 0.000646 ***
...
```

Notice that, once age and height were removed, bmp was no longer significant. In the original reference (Altman, 1991), weight, fev1, and bmp all ended up with p-values below 5%. However, far from all elimination procedures lead to that result.

It is also a good idea to pay close attention to the age, weight, and height variables, which are heavily correlated since we are dealing with children and adolescents.

```
> summary(lm(pemax~age+weight+height))
...
              Estimate Std. Error t value Pr(>|t|)
(Intercept)  64.65555   82.40935    0.785    0.441
age           1.56755    3.14363    0.499    0.623
weight        0.86949    0.85922    1.012    0.323
height       -0.07608    0.80278   -0.095    0.925
...
> summary(lm(pemax~age+height))
...
              Estimate Std. Error t value Pr(>|t|)
(Intercept)   17.8600    68.2493    0.262    0.796
```

```
age                 2.7178      2.9325    0.927     0.364
height              0.3397      0.6900    0.492     0.627
...
> summary(lm(pemax~age))
...
              Estimate Std. Error t value Pr(>|t|)
(Intercept)     50.408      16.657   3.026  0.00601 **
age              4.055       1.088   3.726  0.00111 **
...
> summary(lm(pemax~height))
...
              Estimate Std. Error t value Pr(>|t|)
(Intercept) -33.2757      40.0445  -0.831  0.41453
height        0.9319       0.2596   3.590  0.00155 **
...
```

As it turns out, there is really no reason to prefer one of the three variables over the two others. The fact that an elimination method ends up with a model containing only weight is essentially a coincidence. You can easily be misled by model search procedures that end up with one highly significant variable — it is far from certain that the same variable would be chosen if you were to repeat the analysis on a new, similar data set.

What you may reasonably conclude is that there is probably a connection with the patient's physical development or size, which may be described in terms of age, height, or weight. Which description to use is arbitrary. If you want to choose one over the others, a decision cannot be based on the data, although possibly on theoretical considerations and/or results from previous investigations.

11.4 Exercises

11.1 The secher data are best analyzed after log-transforming birth weight as well as the abdominal and biparietal diameters. Fit a prediction equation for birth weight. How much is gained by using both diameters in a prediction equation? The sum of the two regression coefficients is almost exactly 3 — can this be given a nice interpretation?

11.2 The tlc data set contains a variable also called tlc. This is not in general a good idea; explain why. Describe tlc using the other variables in the data set and discuss the validity of the model.

11.3 The analyses of cystfibr involve sex, which is a binary variable. How would you interpret the results for this variable?

11.4 Consider the juul2 data set and select the group of those over 25 years old. Perform a regression analysis of $\sqrt{\text{igf1}}$ on age, and extend the model by including `height` and `weight`. Generate the analysis of variance table for the extended model. What is the surprise, and why does it happen?

11.5 Analyze and interpret the effect of explanatory variables on the milk intake in the `kfm` data set using a multiple regression model. Notice that `sex` is a factor here; what does that imply for the analyses?

12

Linear models

Many data sets are inherently too complex to be handled adequately by standard procedures and thus require the formulation of ad hoc models. The class of *linear models* provides a flexible framework into which many — although not all — of these cases can be fitted.

You may have noticed that the lm function is applied to data classified into groups (Chapter 7) as well as to (multiple) linear regression (Chapters 6 and 11) problems, even though the theory for these procedures appears to be quite different. However, they are, in fact, special cases of the same general model.

The basic point is that a multiple regression model can describe a wide variety of situations if you choose the explanatory variables suitably. There is no requirement that the explanatory variables should follow a normal distribution, or any continuous distribution for that matter. One simple example (which we use without comment in Chapter 11) is that a grouping into two categories can be coded as a 0/1 variable and used in a regression analysis. The regression coefficient in that case corresponds to a difference between two groups rather than the slope of an actual line. To encode a grouping with more than two categories, you can use multiple 0/1 variables.

Generating these *dummy variables* becomes tedious, but it can be automated by the use of model formulas. Among other things, such formulas provide a convenient abstraction by treating classification variables (factors) and continuous variables symmetrically. You will need to learn

P. Dalgaard, *Introductory Statistics with R*,
DOI: 10.1007/978-0-387-79054-1_12, © Springer Science+Business Media, LLC 2008

exactly what model formulas do in order to become able to express your own modelling ideas.

This chapter contains a collection of models and their handling by `lm`, mainly in the form of relatively minor extensions and modifications of methods described earlier. It is meant only to give you a feel for the scope of possibilities and does not pretend to be complete.

12.1   Polynomial regression

One basic observation showing that multiple regression analysis can do more than meets the eye is that you can include second-order and higher powers of a variable in the model along with the original linear term. That is, you can have a model like

$$y = \alpha + \beta_1 x + \beta_2 x^2 + \cdots + \beta_k x^k + \epsilon$$

This obviously describes a nonlinear relation between y and x, but that does not matter; the model is still a linear model. What does matter is that the relation between the *parameters* and the expected observations is linear. It also does not matter that there is a deterministic relation between the regression variables $x, x^2, x^3, \ldots$, as long as there is no *linear* relation between them. However, fitting high-degree polynomials can be difficult because near-collinearity between terms makes the fit numerically unstable.

We return to the cystic fibrosis data set for an example. The plot of `pemax` and `height` in Figure 11.1 may suggest that the relation is not quite linear. One way to test this is to try to add a term that is the square of the height.

```
> attach(cystfibr)
> summary(lm(pemax~height+I(height^2)))
...
             Estimate Std. Error t value Pr(>|t|)
(Intercept) 615.36248  240.95580   2.554   0.0181 *
height        -8.08324    3.32052  -2.434   0.0235 *
I(height^2)    0.03064    0.01126   2.721   0.0125 *
...
```

Notice that the computed height^2 in the model formula needs to be "protected" by `I(...)`. This technique is often used to prevent any special interpretation of operators in a model formula. Such an interpretation will not take place inside a function call, and `I` is the *identity* function that returns its argument unaltered.

We find a significant deviation from linearity. However, considering the process that led to doing this particular analysis, the *p*-values have to be taken with more than a grain of salt. This is getting dangerously close to "data dredging", fishing expeditions in data. Consider it more an illustration of a technique than an exemplary data analysis.

To draw a plot of the fitted curve with prediction and confidence bands, we can use `predict`. To avoid problems caused by data not being sorted by height, we use `newdata`, which allows the prediction of values for a chosen set of predictors. Here we choose a set of heights between 110 and 180 cm in steps of 2 cm:

```
> pred.frame <- data.frame(height=seq(110,180,2))
> lm.pemax.hq <- lm(pemax~height+I(height^2))
> predict(lm.pemax.hq,interval="pred",newdata=pred.frame)
          fit      lwr      upr
1    96.90026 37.94461 155.8559
2    94.33611 36.82985 151.8424
3    92.01705 35.73077 148.3033
...
34 141.68922 88.70229 194.6761
35 147.21294 93.51117 200.9147
36 152.98174 98.36718 207.5963
```

Based on these predicted data, Figure 12.1 is obtained as follows:

```
> pp <- predict(lm.pemax.hq,newdata=pred.frame,interval="pred")
> pc <- predict(lm.pemax.hq,newdata=pred.frame,interval="conf")
> plot(height,pemax,ylim=c(0,200))
> matlines(pred.frame$height,pp,lty=c(1,2,2),col="black")
> matlines(pred.frame$height,pc,lty=c(1,3,3),col="black")
```

It is seen that the fitted curve is slightly decreasing for small heights. This is probably an artifact caused by the choice of a second-order polynomial to fit data. More likely, the reality is that `pemax` is relatively constant up to about 150 cm, after which it increases quickly with height. Note also that there seems to be a discrepancy between the prediction limits and the actual distribution of data for the smaller heights. The standard deviation might be larger for larger heights, but it is not impossible to obtain a similar distribution of points by coincidence, and there is also an issue with potential overfitting to the observed data. It is really not advisable to construct prediction intervals based on data as limited as these unless you are sure that the model is correct.

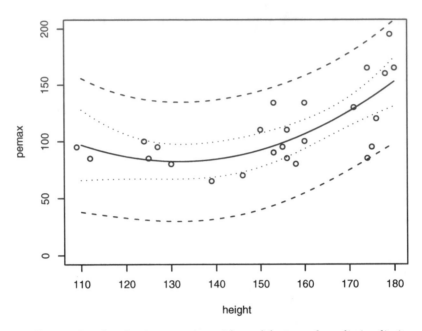

Figure 12.1. Quadratic regression with confidence and prediction limits.

12.2   Regression through the origin

It sometimes makes sense to assume that a regression line passes through $(0,0)$ — that the intercept of the regression line is zero. This can be specified in the model formula by adding the term `-1` ("minus intercept") to the right-hand side: `y ~ x - 1`.

The logic of the notation can be seen by writing the linear regression model as $y = \alpha \times 1 + \beta \times x + \epsilon$. The intercept corresponds to having an extra descriptive variable, which is the constant 1. Removing this variable yields regression through the origin.

This is a simulated example of a linear relationship through the origin $(y = 2x + \epsilon)$:

```
> x <- runif(20)
> y <- 2*x+rnorm(20,0,0.3)
> summary(lm(y~x))

Call:
lm(formula = y ~ x)
```

```
Residuals:
     Min      1Q    Median      3Q      Max
-0.50769 -0.08766  0.03802  0.14512  0.26358

Coefficients:
            Estimate Std. Error t value Pr(>|t|)
(Intercept) -0.14896    0.08812   -1.69    0.108
x            2.39772    0.15420   15.55 7.05e-12 ***
---
Signif. codes:  0 '***' 0.001 '**' 0.01 '*' 0.05 '.' 0.1 ' ' 1

Residual standard error: 0.2115 on 18 degrees of freedom
Multiple R-squared: 0.9307,    Adjusted R-squared: 0.9269
F-statistic: 241.8 on 1 and 18 DF,  p-value: 7.047e-12


> summary(lm(y~x-1))

Call:
lm(formula = y ~ x - 1)

Residuals:
     Min      1Q    Median      3Q      Max
-0.62178 -0.16855 -0.04019  0.12044  0.27346

Coefficients:
  Estimate Std. Error t value Pr(>|t|)
x  2.17778    0.08669   25.12 4.87e-16 ***
---
Signif. codes:  0 '***' 0.001 '**' 0.01 '*' 0.05 '.' 0.1 ' ' 1

Residual standard error: 0.2216 on 19 degrees of freedom
Multiple R-squared: 0.9708,    Adjusted R-squared: 0.9692
F-statistic: 631.1 on 1 and 19 DF,  p-value: 4.873e-16
```

In the first analysis, the intercept is not significant, which is, of course, not surprising. In the second analysis we force the intercept to be zero, resulting in a slope estimate with a substantially improved accuracy.

Comparison of the R^2-values in the two analyses shows something that occasionally causes confusion: R^2 is much larger in the model with no intercept! This does *not*, however, mean that the relation is "more linear" when the intercept is not included or that more of the variation is explained. What is happening is that the definition of R^2 itself is changing. It is most easily seen from the ANOVA tables in the two cases:

```
> anova(lm(y~x))
Analysis of Variance Table

Response: y
          Df  Sum Sq Mean Sq F value    Pr(>F)
x          1 10.8134 10.8134  241.80 7.047e-12 ***
Residuals 18  0.8050  0.0447
```

```
---
Signif. codes:  0 '***' 0.001 '**' 0.01 '*' 0.05 '.' 0.1 ' ' 1
> anova(lm(y~x-1))
Analysis of Variance Table

Response: y
          Df  Sum Sq Mean Sq F value      Pr(>F)
x          1 30.9804 30.9804  631.06 4.873e-16 ***
Residuals 19  0.9328  0.0491
---
Signif. codes:  0 '***' 0.001 '**' 0.01 '*' 0.05 '.' 0.1 ' ' 1
```

Notice that the total sum of squares and the total number of degrees of freedom is not the same in the two analyses. In the model with an intercept, there are 19 DF in all and the total sum of squares is $\sum(y_i - \bar{y})^2$, while the model without an intercept has a total of 20 DF and the total sum of squares is defined as $\sum y_i^2$. Unless $\bar{y}$ is close to zero, the latter "total SS" will be much larger than the former, so if the residual variance is similar, R^2 will be much closer to 1.

The reason for defining the total sum of squares like this for models without intercepts is that it has to correspond to the residual sum of squares in a minimal model. The minimal model has to be a submodel of the regression model; otherwise the ANOVA table simply does not make sense. In an ordinary regression analysis, the minimal model is $y = \alpha + \epsilon$, but when the regression model does not include α, the only sensible minimal model is $y = 0 + \epsilon$.

12.3   Design matrices and dummy variables

The function model.matrix gives the *design matrix* for a given model. It can look like this:

```
> model.matrix(pemax~height+weight)
   (Intercept) height weight
1            1    109   13.1
2            1    112   12.9
3            1    124   14.1
4            1    125   16.2
...
24           1    175   51.1
25           1    179   71.5
attr(,"assign")
[1] 0 1 2
```

(The cystfibr data set was attached previously.)

You should not worry about the "assign" attribute at this stage, but the three columns are important. If you add them together, weighted by the corresponding regression coefficients, you get exactly the fitted values. Notice that the intercept enters as the coefficient to a column of ones.

If the same is attempted for a model containing a factor, the following happens. (We return to the anesthetic ventilation example from p. 129.)

```
> attach(red.cell.folate)
> model.matrix(folate~ventilation)
   (Intercept) ventilationN2O+O2,op ventilationO2,24h
1            1                    0                 0
2            1                    0                 0
...
16           1                    1                 0
17           1                    1                 0
18           1                    0                 1
19           1                    0                 1
20           1                    0                 1
21           1                    0                 1
22           1                    0                 1
attr(,"assign")
[1] 0 1 1
attr(,"contrasts")
attr(,"contrasts")$ventilation
[1] "contr.treatment"
```

The two columns of zeros and ones are sometimes called *dummy variables.* They are interpreted exactly as above: Multiplying them by the respective regression coefficients and adding the results yields the fitted value. Notice that, for example, the second column is 1 for observations in group 2 and 0 otherwise; that is, the corresponding regression coefficient describes something that is added to the intercept for observations in that particular group. Both columns have zeros for observations from the first group, the mean value of which is described by the intercept (β_0) alone. The regression coefficient β_1 thus describes the *difference* in means between groups 1 and 2, and β_2 between groups 1 and 3.

You may be confused by the use of the term "regression coefficients" even though no regression lines are present in models like that above. The point is that you *formally* rewrite a model for groups as a multiple regression model, so that you can use the same software. As is seen, there is a unique correspondence between the formal regression coefficients and the group means.

You can define dummy variables in several different ways to describe a grouping. This particular scheme is called *treatment contrasts* because if the first group is "no treatment" then the coefficients immediately give the treatment effects for each of the other groups. We do not discuss other choices here; see Venables and Ripley (2002) for a much deeper discussion. Note only that contrast type can be set on a per-term basis and that this is what is reflected in the "contrasts" attribute of the design matrix.

For completeness, the "assign" attribute indicates which columns belong together. When, for instance, you request an analysis of variance using anova, the sum of squares for ventilation will have 2 degrees of freedom, corresponding to the removal of both columns simultaneously.

Removing the intercept from a model containing a factor term will not correspond to a model in which a particular group has mean zero since such models are usually nonsensical. Instead, R generates a simpler set of dummy variables, which are indicator variables of the levels of the factor. This corresponds to the same model as when the intercept is included (the fitted values are identical), but the regression coefficients have a different interpretation.

12.4   Linearity over groups

Sometimes data are grouped according to a division of a continuous scale (e.g., by age group), or an experiment was designed to take several measurements at each of a fixed set of x-values. In both cases it is relevant to compare the results of a linear regression with those of an analysis of variance.

In the case of grouped x-values, you might take a central value as representative for everyone in a given group, for instance formally letting everyone in a "20–29-year" category be 25 years old. If individual x-values are available, they may of course be used in a linear regression, but it makes the analysis a little more complicated, so we discuss only the situation where that is not the case.

We thus have two alternative models for the same data. Both belong to the class of linear models that lm is capable of handling. The linear regression model is a *submodel* of the model for one-way analysis of variance because the former can be obtained by placing restrictions on the parameters of the latter (namely that the true group means lie on a straight line).

It is possible to test whether or not a model reduction is allowable by comparing the reduction in the amount of variation explained to the residual variation in the larger model, resulting in an F test.

In the following example on trypsin concentrations in age groups (Altman, 1991, p. 212), data are given as the mean and SD within each of six groups. This is a kind of data that R is not quite prepared to handle, and it has therefore been necessary to create "fake" data giving the same means and SDs. These can be obtained via

```
> attach(fake.trypsin)
```

The actual results of the analysis of variance depend only on the means and SDs and are therefore independent of the faking. Readers interested in how to perform the actual faking should take a look at the file fake.trypsin.R in the rawdata directory of the ISwR package.

The fake.trypsin data frame contains three variables, as seen by

```
> summary(fake.trypsin)
    trypsin              grp            grpf
 Min.   :-39.96   Min.   :1.000   1: 32
 1st Qu.:119.52   1st Qu.:2.000   2:137
 Median :167.59   Median :2.000   3: 38
 Mean   :168.68   Mean   :2.583   4: 44
 3rd Qu.:213.98   3rd Qu.:3.000   5: 16
 Max.   :390.13   Max.   :6.000   6:  4
```

Notice that there are both grp, which is a numeric vector, and grpf, which is a factor with six levels.

Performing a one-way analysis of variance on the fake data gives the following ANOVA table:

```
> anova(lm(trypsin~grpf))
Analysis of Variance Table

Response: trypsin
           Df Sum Sq Mean Sq F value    Pr(>F)
grpf        5 224103   44821  13.508 9.592e-12 ***
Residuals 265 879272    3318
```

If you had used grp instead of grpf in the model formula, you would have obtained a linear regression on the group number instead. In some circumstances, that would have been a serious error, but here it actually makes sense. The midpoints of the age intervals are equidistant, so the model is equivalent to assuming a linear development with age (the interpretation of the regression coefficient requires some care, though). The ANOVA table looks as follows:

```
> anova(lm(trypsin~grp))
Analysis of Variance Table

Response: trypsin
           Df Sum Sq Mean Sq F value    Pr(>F)
grp         1 206698  206698  62.009 8.451e-14 ***
Residuals 269 896677    3333
```

Notice that the residual mean squares did not change very much, indicating that the two models describe the data nearly equally well. If you want to have a formal test of the simple linear model against the model where there is a separate mean for each group, it can be done easily as follows:

```
> model1 <- lm(trypsin~grp)
> model2 <- lm(trypsin~grpf)
> anova(model1,model2)
Analysis of Variance Table

Model 1: trypsin ~ grp
Model 2: trypsin ~ grpf
  Res.Df    RSS  Df Sum of Sq      F Pr(>F)
1    269 896677
2    265 879272   4     17405 1.3114 0.2661
```

So we see that the model reduction has a nonsignificant *p*-value and hence that model2 does not fit data significantly better than model1.

This technique works *only* when one model is a submodel of the other, which is the case here since the linear model is defined by a restriction on the group means.

Another way to achieve the same result is to add the two models together formally as follows:

```
> anova(lm(trypsin~grp+grpf))
Analysis of Variance Table

Response: trypsin
           Df Sum Sq Mean Sq F value    Pr(>F)
grp         1 206698  206698 62.2959 7.833e-14 ***
grpf        4  17405    4351  1.3114    0.2661
Residuals 265 879272    3318
```

This model is exactly the same as when only grpf was included. However, the ANOVA table now contains a subdivision of the model sum of squares in which the grpf line describes the *change* incurred by expanding the model from one to five parameters. The ANOVA table in Altman (1991, p. 213) is different, erroneously.

The plot in Figure 12.2 is made like this:

```
> xbar.trypsin <- tapply(trypsin,grpf,mean)
> stripchart(trypsin~grp, method="jitter",
+    jitter=.1, vertical=T, pch=20)
> lines(1:6,xbar.trypsin,type="b",pch=4,cex=2,lty=2)
> abline(lm(trypsin~grp))
```

The graphical techniques used here are essentially identical to those used for Figure 7.1, so we do not go into further details.

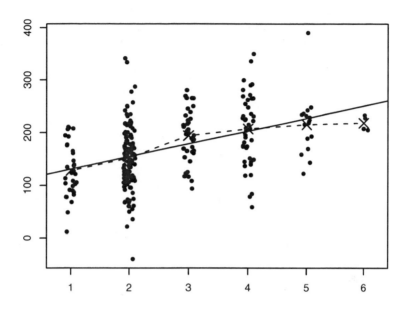

Figure 12.2. "Fake" data for the trypsin example with fitted line and empirical means.

Notice that the fakeness of the data is exposed by a point showing a negative trypsin concentration! The original data are unavailable but would likely show a distribution skewed slightly upwards.

Actually, it *is* possible to analyze the data in R without generating fake data. A weighted regression analysis of the group means, with weights equal to the number of observations in each group, will yield the first two lines of the ANOVA table, and the last one can be computed from the SDs. The details are as follows:

```
> n <- c(32,137, 38,44,16,4)
> tryp.mean <- c(128,152,194,207,215,218)
> tryp.sd <-c(50.9,58.5,49.3,66.3,60,14)
> gr<-1:6
> anova(lm(tryp.mean~gr+factor(gr),weights=n))
Analysis of Variance Table

Response: tryp.mean
           Df Sum Sq Mean Sq F value Pr(>F)
gr          1 206698  206698
factor(gr)  4  17405    4351
Residuals   0      0
```

Notice that the "Residuals" line is zero and that the F tests are not calculated. Omitting the `factor(gr)` term will cause that line to go into Residuals and be treated as an estimate of the error variation, but that is not what you want since it does not include the information about the variation within groups. Instead, you need to fill in the missing information computed from the group standard deviations and sizes. The following gives the residual sum of squares and the corresponding degrees of freedom and mean squares:

```
> sum(tryp.sd^2*(n-1))
[1] 879271.9
> sum(n-1)
[1] 265
> sum(tryp.sd^2*(n-1))/sum(n-1)
[1] 3318.007
```

There is no simple way of updating the ANOVA table with an external variance estimate, but it is easy enough to do the computations directly:

```
> 206698/3318.007 # F statistic for gr
[1] 62.29583           .
> 1-pf(206698/3318.007,1,265) # p-value
[1] 7.838175e-14
> 4351/3318.007   # F statistic for factor(gr)
[1] 1.311329
> 1-pf(4351/3318.007,4,265) # p-value
[1] 0.2660733
```

12.5   Interactions

A basic assumption in a multiple regression model is that terms act additively on the response. However, this does not mean that linear models cannot describe nonadditivity. You can add special *interaction terms* that specify that the effect of one term is modified according to the level of another. In the model formulas in R, such terms are generated using the colon operator; for example, a:b. Usually, you will also include the terms a and b, and R allows the notation a*b for a+b+a:b. Higher-order interactions among three or more variables are also possible.

The exact definition of the interaction terms and the interpretation of their associated regression coefficients can be elusive. Some peculiar things happen if an interaction term is present but one or more of the main effects are missing. The full details are probably best revealed through experimentation. However, depending on the nature of the terms a and b as factors or numeric variables, the overall effect of including interaction terms can be described as follows:

- *Interaction between two factors.* This is conceptually the simplest case. The model with interaction corresponds to having different levels for all possible combinations of levels of the two factors.

- *Interaction between a factor and a numeric variable.* In this case, the model with interaction contains linear effects of the continuous variable but with different slopes within each group defined by the factor.

- *Interaction between two continuous variables.* This gives a slightly peculiar model containing a new regression variable that is the product of the two. The interpretation is that you have a linear effect of varying one variable while keeping the other constant, but with a slope that changes as you vary the other variable.

12.6   Two-way ANOVA with replication

The `coking` data set comes from Johnson (1994, Section 13.1). The time required to make coke from coal is analyzed in a 2×3 experiment varying the oven temperature and the oven width. There were three replications at each combination.

```
> attach(coking)
> anova(lm(time~width*temp))
Analysis of Variance Table

Response: time
            Df  Sum Sq Mean Sq F value     Pr(>F)
width        2 123.143  61.572 222.102 3.312e-10 ***
temp         1  17.209  17.209  62.076 4.394e-06 ***
width:temp   2   5.701   2.851  10.283  0.002504 **
Residuals   12   3.327   0.277
---
Signif. codes:  0 '***' 0.001 '**' 0.01 '*' 0.05 '.' 0.1 ' ' 1
```

We see that the interaction term is significant. If we take a look at the cell means, we can get an idea of why this happens:

```
> tapply(time,list(width,temp),mean)
          1600      1900
4     3.066667 2.300000
8     7.166667 5.533333
12 10.800000 7.333333
```

The difference between high and low temperatures increases with oven width, making an additive model inadequate. When this is the case, the individual tests for the two factors make no sense. If the interaction had not been significant, then we would have been able to perform separate F tests for the two factors.

12.7   Analysis of covariance

As the example in this section, we use a data set concerning growth conditions of *Tetrahymena* cells, collected by Per Hellung-Larsen. Data are from two groups of cell cultures where glucose was either added or not added to the growth medium. For each culture, the average cell diameter (μ) and cell concentration (count per ml) were recorded. The cell concentration was set at the beginning of the experiment, and there is no systematic difference in cell concentration between the two glucose groups. However, it is expected that the cell diameter is affected by the presence of glucose in the medium.

Data are in the data frame hellung, which can be loaded and viewed like this:

```
> hellung
    glucose    conc diameter
1         1  631000     21.2
2         1  592000     21.5
3         1  563000     21.3
4         1  475000     21.0
...
49        2   14000     24.4
50        2   13000     24.3
51        2   11000     24.2
```

The coding of glucose is such that 1 and 2 mean yes and no, respectively. There are no missing values.

Summarizing the data frame yields

```
> summary(hellung)
    glucose          conc              diameter
 Min.   :1.000   Min.   : 11000    Min.   :19.20
 1st Qu.:1.000   1st Qu.: 27500    1st Qu.:21.40
 Median :1.000   Median : 69000    Median :23.30
 Mean   :1.373   Mean   :164325    Mean   :23.00
 3rd Qu.:2.000   3rd Qu.:243000    3rd Qu.:24.35
 Max.   :2.000   Max.   :631000    Max.   :26.30
```

Notice that the distribution of the concentrations is strongly right-skewed with a mean more than twice as big as the median. Note also that glucose is regarded as a numeric vector by summary, even though it has only two different values.

It will be more convenient to have `glucose` as a factor, so it is recoded as shown below. Recall that to change a variable inside a data frame, you use `$`-notation (p. 21) to specify the component you want to change:

```
> hellung$glucose <- factor(hellung$glucose, labels=c("Yes","No"))
> summary(hellung)
   glucose        conc            diameter
 Yes:32    Min.   : 11000   Min.   :19.20
 No :19    1st Qu.: 27500   1st Qu.:21.40
           Median : 69000   Median :23.30
           Mean   :164325   Mean   :23.00
           3rd Qu.:243000   3rd Qu.:24.35
           Max.   :631000   Max.   :26.30
```

It is convenient to be able to refer to the variables of `hellung` without the `hellung$` prefix, so we put `hellung` in the search path.

```
> attach(hellung)
```

12.7.1 Graphical description

First, we plot the raw data (Figure 12.3):

```
> plot(conc,diameter,pch=as.numeric(glucose))
```

By calculating `as.numeric(glucose)`, we convert the factor `glucose` to the underlying codes, 1 and 2. The specification of `pch` thus implies that group 1 ("Yes") is drawn using plotting character 1 (circles) and group 2 with plotting character 2 (triangles).

To get different plotting symbols, you must first create a vector containing the symbol numbers and give that as the `pch` argument. The following form yields open and filled circles: `c(1,16)[glucose]`. It looks a bit cryptic at first, but it is really just a consequence of R's way of indexing. For indexing purposes, a factor like `glucose` behaves as a vector of 1s and 2s, so you get the first element of `c(1,16)`, namely 1, whenever an observation is from group 1; when the observation is from group 2, you similarly get 16.

The explanatory text is inserted with `legend` like this:

```
> legend(locator(n=1),legend=c("glucose","no glucose"),pch=1:2)
```

Notice that both the function and one of its arguments are named `legend`.

The function `locator` returns the coordinates of a point on a plot. It works so that the function awaits a click with a mouse button and then returns the cursor position. You may want to call `locator()` directly from

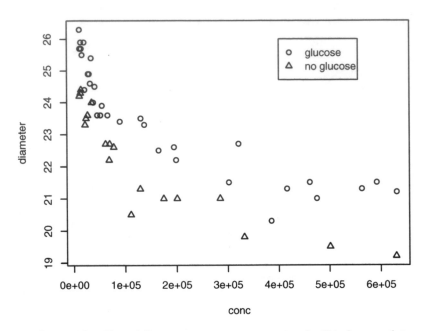

Figure 12.3. Plot of diameter versus concentration for *Tetrahymena* data.

the command line to see the effect. Notice that if you do not specify a value for n, then you need to right-click when you are done selecting points.

The plot shows a clear inverse and nonlinear relation between concentration and cell diameter. Further, it is seen that the cultures without glucose are systematically below cultures with added glucose.

You get a much nicer plot (Figure 12.4) by using a logarithmic *x*-axis:

```
> plot(conc,diameter,pch=as.numeric(glucose),log="x")
```

Now the relation suddenly looks linear!

You could also try a log-log plot (shown in Figure 12.5 with regression lines as described below):

```
> plot(conc,diameter,pch=as.numeric(glucose),log="xy")
```

As is seen, this really does not change much, but it was nevertheless decided to analyze data with both diameter and concentration log-transformed because a power-law relation was expected ($y = \alpha x^\beta$, which gives a straight line on a log-log plot).

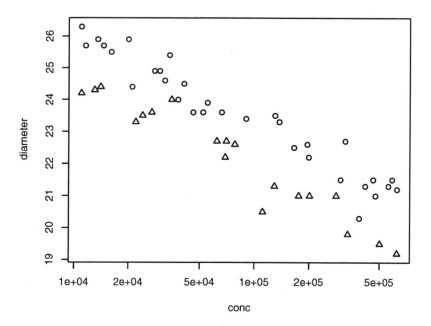

Figure 12.4. *Tetrahymena* data with logarithmic *x*-axis.

When adding regression lines to a log plot or log-log plot, you should notice that `abline` interprets them as lines in the coordinate system obtained *after* taking (base-10) logarithms. Thus, you can add a line for each group with `abline` applied to the results of a regression analysis of `log10(diameter)` on `log10(conc)`. First, however, it is convenient to define data frames corresponding to the two glucose groups:

```
> tethym.gluc <- hellung[glucose=="Yes",]
> tethym.nogluc <- hellung[glucose=="No",]
```

Notice that you have to use the names, not the numbers, of the factor levels.

Since we only need the two data frames for adding lines to the figure, it would be cumbersome to add them in turn to the search path with `attach`, do the plotting, and then use `detach` to remove them. It is easier to use the `data` argument to `lm`; this allows you to explicitly specify the data frame in which to look for variables. The two regression lines are drawn with

```
> lm.nogluc <- lm(log10(diameter) ~ log10(conc),data=tethym.nogluc)
> lm.gluc <- lm(log10(diameter) ~ log10(conc),data=tethym.gluc)
```

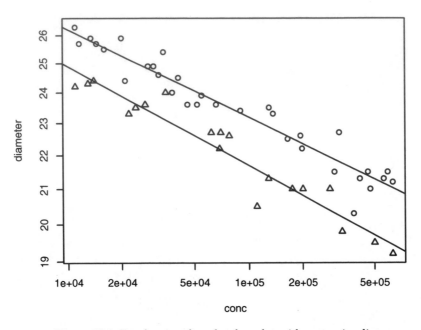

Figure 12.5. *Tetrahymena* data, log-log plot with regression lines.

```
> abline(lm.nogluc)
> abline(lm.gluc)
```

after which the plot looks like Figure 12.5. It is seen that the lines fit the data quite well and that they are almost, but not perfectly, parallel. The question is whether the difference in slope is statistically significant. This is the topic of the next section.

12.7.2   *Comparison of regression lines*

Corresponding to the two lines from before, we have the following regression analyses:

```
> summary(lm(log10(diameter)~ log10(conc), data=tethym.gluc))

Call:
lm(formula = log10(diameter) ~ log10(conc), data = tethym.gluc)

Residuals:
        Min         1Q      Median          3Q         Max
-0.0267219  -0.0043361   0.0006891   0.0035489   0.0176077
```

```
Coefficients:
           Estimate Std. Error t value Pr(>|t|)
(Intercept)  1.63134    0.01345  121.29   <2e-16 ***
log10(conc) -0.05320    0.00272  -19.56   <2e-16 ***
---
Signif. codes:  0 '***' 0.001 '**' 0.01 '*' 0.05 '.' 0.1 ' ' 1

Residual standard error: 0.008779 on 30 degrees of freedom
Multiple R-squared: 0.9273,     Adjusted R-squared: 0.9248
F-statistic: 382.5 on 1 and 30 DF,   p-value: < 2.2e-16

> summary(lm(log10(diameter)~ log10(conc), data=tethym.nogluc))

Call:
lm(formula = log10(diameter) ~ log10(conc), data = tethym.nogluc)

Residuals:
      Min         1Q     Median         3Q        Max
-2.192e-02 -4.977e-03  5.598e-05  5.597e-03  1.663e-02

Coefficients:
           Estimate Std. Error t value Pr(>|t|)
(Intercept)  1.634761   0.020209   80.89  < 2e-16 ***
log10(conc) -0.059677   0.004125  -14.47 5.48e-11 ***
---
Signif. codes:  0 '***' 0.001 '**' 0.01 '*' 0.05 '.' 0.1 ' ' 1

Residual standard error: 0.009532 on 17 degrees of freedom
Multiple R-squared: 0.9249,     Adjusted R-squared: 0.9205
F-statistic: 209.3 on 1 and 17 DF,   p-value: 5.482e-11
```

Notice that you can use arithmetic expressions in the model formula [here log10(...)]. There are limitations, though, because, for example, z~x+y means a model where z is described by an additive model in x and y, which is not the same as a regression analysis on the sum of the two. The latter may be specified using z~I(x+y) (I for "identity").

A quick assessment of the significance of the difference between the slopes of the two lines can be obtained as follows: The difference between the slope estimates is 0.0065, and the standard error of that is $\sqrt{0.0041^2 + 0.0027^2} = 0.0049$. Since $t = 0.0065/0.0049 = 1.3$, it would seem that we are allowed to assume that the slopes are the same.

It is, however, preferable to fit a model to the entire data set and test the hypothesis of equal slopes in that model. One reason that this approach is preferable is that it can be generalized to more complicated models. Another reason is that even though there is nothing seriously wrong with the simple test for equal slopes, that procedure gives you little information on how to proceed. If the slopes are the same, you would naturally want to find an estimate of the common slope and the distance between the parallel lines.

First, we set up a model that allows the relation between concentration and cell diameter to have different slopes and intercepts in the two glucose groups:

```
> summary(lm(log10(diameter)~log10(conc)*glucose))

Call:
lm(formula = log10(diameter) ~ log10(conc) * glucose)

Residuals:
      Min         1Q     Median         3Q        Max
-2.672e-02 -4.888e-03  5.598e-05  3.767e-03  1.761e-02

Coefficients:
                          Estimate Std. Error t value Pr(>|t|)
(Intercept)               1.631344   0.013879 117.543  <2e-16 ***
log10(conc)              -0.053196   0.002807 -18.954  <2e-16 ***
glucoseNo                 0.003418   0.023695   0.144   0.886
log10(conc):glucoseNo    -0.006480   0.004821  -1.344   0.185
---
Signif. codes:  0 '***' 0.001 '**' 0.01 '*' 0.05 '.' 0.1 ' ' 1

Residual standard error: 0.009059 on 47 degrees of freedom
Multiple R-squared: 0.9361,    Adjusted R-squared: 0.9321
F-statistic: 229.6 on 3 and 47 DF,  p-value: < 2.2e-16
```

These regression coefficients should be read as follows. The expected value of the log cell diameter for an observation with cell concentration C is obtained as the sum of the following four quantities:

1. The intercept, 1.6313
2. $-0.0532 \times \log_{10} C$
3. 0.0034, but only for a culture without glucose
4. $-0.0065 \times \log_{10} C$, but only for cultures without glucose

Accordingly, for cell cultures with glucose, we have the linear relation

$$\log_{10} D = 1.6313 - 0.0532 \times \log_{10} C$$

and for cultures without glucose we have

$$\log_{10} D = (1.6313 + 0.0034) - (0.0532 + 0.0065) \times \log_{10} C$$

Put differently, the first two coefficients in the joint model can be interpreted as the estimates for intercept and slope in group 1, whereas the latter two are the differences between group 1 and group 2 in intercept and slope, respectively. Comparison with the separate regression analyses shows that slopes and intercepts are the same as in the joint analysis. The standard errors differ a little from the separate analyses because a pooled variance estimate is now used. Notice that the rough test of difference in slope outlined above is essentially the *t* test for the last coefficient.

Notice also that the `glucose` and `log10(conc).glucose` terms indicate items to be added for cultures *without* glucose. This is because the factor levels are ordered yes = 1 and no = 2, and the base level is the first group.

Fitting an additive model, we get

```
> summary(lm(log10(diameter)~log10(conc)+glucose))
...
Coefficients:
              Estimate Std. Error t value Pr(>|t|)
(Intercept)   1.642132   0.011417  143.83  < 2e-16 ***
log10(conc)  -0.055393   0.002301  -24.07  < 2e-16 ***
glucoseNo    -0.028238   0.002647  -10.67 2.93e-14 ***
...
```

Here the interpretation of the coefficients is that the estimated relation for cultures with glucose is

$$\log_{10} D = 1.6421 - 0.0554 \times \log_{10} C$$

and for cultures without glucose it is

$$\log_{10} D = (1.6421 - 0.0282) - 0.0554 \times \log_{10} C$$

That is, the lines for the two cultures are parallel, but the log diameters for cultures without glucose are 0.0282 below those with glucose. On the original (nonlogarithmic) scale, this means that the former are 6.3% lower (a constant absolute difference on a logarithmic scale corresponds to constant relative differences on the original scale and $10^{-0.0282} = 0.937$).

The joint analysis presumes that the variance around the regression line is the same in the two groups. This assumption should really have been tested before embarking on the analysis above. A formal test can be performed with `var.test`, which conveniently allows a pair of linear models as arguments instead of a model formula or two group vectors:

```
> var.test(lm.gluc,lm.nogluc)

        F test to compare two variances

data:  lm.gluc and lm.nogluc
F = 0.8482, num df = 30, denom df = 17, p-value = 0.6731
alternative hypothesis: true ratio of variances is not equal to 1
95 percent confidence interval:
 0.3389901 1.9129940
sample estimates:
```

```
ratio of variances
        0.8481674
```

When there are more than two groups, Bartlett's test can be used. It, too, allows linear models to be compared. The reservations about robustness against nonnormality apply here, too.

It is seen that it is possible to assume that the lines have the same slope and that they have the same intercept, but — as we will see below — not both at once. The hypothesis of a common intercept is silly anyway unless the slopes are also identical: The intercept is by definition the y-value at $x = 0$, which because of the log scale corresponds to a cell concentration of 1. That is far outside the region the data cover, and it is a completely arbitrary point that will change if the concentrations are measured in different units.

The ANOVA table for the model is

```
> anova(lm(log10(diameter)~ log10(conc)*glucose))
Analysis of Variance Table

Response: log10(diameter)
                    Df   Sum Sq  Mean Sq F value    Pr(>F)
log10(conc)          1 0.046890 0.046890 571.436 < 2.2e-16 ***
glucose              1 0.009494 0.009494 115.698  2.89e-14 ***
log10(conc):glucose  1 0.000148 0.000148   1.807    0.1853
Residuals           47 0.003857 0.000082
---
Signif. codes:  0 `***' 0.001 `**' 0.01 `*' 0.05 `.' 0.1 ` ' 1
```

The model formula a*b, where in the present case a is log10(conc) and b is glucose, is a short form for a + b + a:b, which is read "effect of a plus effect of b plus interaction". The F test in the penultimate line of the ANOVA table is a test for the hypothesis that the last term (a:b) can be omitted, reducing the model to be additive in log10(conc) and glucose, which corresponds to the parallel regression lines. The F test one line earlier indicates whether you can *subsequently* remove glucose and the one in the first line to whether you can additionally remove log10(conc), leaving an empty model.

Alternatively, you can read the table from top to bottom as adding terms describing more and more of the total sum of squares. To those familiar with the SAS system, this kind of ANOVA table is known as type I sums of squares.

The p-value for log10(conc):glucose can be recognized as that of the t test for the coefficient labelled log10(conc).glucose in the previous output. The F statistic is exactly the square of t as well. However, this is true only because there are just two groups. Had there been three or more, there would have been several regression coefficients and the F test would have tested them all against zero simultaneously, just like when all groups are tested equal in a one-way analysis of variance.

Note that the test for removing `log10(conc)` does not make sense because you would have to remove `glucose` first, which is "forbidden" when `glucose` has a highly significant effect. It makes perfectly good sense to test `log10(conc)` *without* removing `glucose` — which corresponds to testing that the two parallel regression lines can be assumed horizontal — but that test is not found in the ANOVA table. You can get the right test by changing the order of terms in the model formula; compare, for instance, these two regression analyses:

```
> anova(lm(log10(diameter)~glucose+log10(conc)))
Analysis of Variance Table

Response: log10(diameter)
            Df   Sum Sq  Mean Sq F value    Pr(>F)
glucose      1 0.008033 0.008033  96.278 4.696e-13 ***
log10(conc)  1 0.048351 0.048351 579.494 < 2.2e-16 ***
Residuals   48 0.004005 0.000083
---
Signif. codes:  0 '***' 0.001 '**' 0.01 '*' 0.05 '.' 0.1 ' ' 1
> anova(lm(log10(diameter)~log10(conc)+ glucose))
Analysis of Variance Table

Response: log10(diameter)
            Df   Sum Sq  Mean Sq F value    Pr(>F)
log10(conc)  1 0.046890 0.046890 561.99 < 2.2e-16 ***
glucose      1 0.009494 0.009494 113.78 2.932e-14 ***
Residuals   48 0.004005 0.000083
---
Signif. codes:  0 '***' 0.001 '**' 0.01 '*' 0.05 '.' 0.1 ' ' 1
```

They both describe exactly the same model, as is indicated by the residual sum of squares being identical. The partitioning of the sum of squares is *not* the same, though — and the difference may be much more dramatic than it is here. The difference is whether `log10(conc)` is added to a model already containing `glucose` or vice versa. Since the second F test in both tables is highly significant, no model reduction is possible and the F test in the line above it is irrelevant.

If you go back and look at the regression coefficients in the model with parallel regression lines, you will see that the squares of the t tests are 579.49 and 113.8, precisely the last F test in the two tables above.

It is informative to compare the covariance analysis above with the simpler analysis in which the effect of cell concentration is ignored:

```
> t.test(log10(diameter)~glucose)
```

```
        Welch Two Sample t-test

data:   log10(diameter) by glucose
t = 2.7037, df = 36.31, p-value = 0.01037
alternative hypothesis: true difference in means is not equal to 0
95 percent confidence interval:
 0.006492194 0.045424241
sample estimates:
mean in group Yes  mean in group No
          1.370046          1.344088
```

Notice that the *p*-value is much less extreme. It is still significant in this case, but in smaller data sets the statistical significance could easily disappear completely. The difference in means between the two groups is 0.026, which is comparable to the 0.028 that was the glucose effect in the analysis of covariance. However, the confidence interval goes from 0.006 to 0.045, whereas the analysis of covariance had 0.023 to 0.034 [$0.0282 \pm t_{.975}(48) \times 0.0026$], which is almost four times as narrow, obviously a substantial gain in efficiency.

12.8   Diagnostics

Regression diagnostics are used to evaluate the model assumptions and investigate whether or not there are observations with a large influence on the analysis. A basic set of these is available via the plot method for lm objects. Four of them are displayed in a 2 × 2 layout (Figure 12.6) as follows:

```
> attach(thuesen)
> options(na.action="na.exclude")
> lm.velo <- lm(short.velocity~blood.glucose)
> opar <- par(mfrow=c(2,2), mex=0.6, mar=c(4,4,3,2)+.3)
> plot(lm.velo, which=1:4)
> par(opar)
```

The par commands set up for a 2 × 2 layout with compressed margin texts and go back to normal after plotting.

The top left panel shows residuals versus fitted values. The top right panel is a Q–Q normal distribution plot of standardized residuals. Notice that there are residuals and standardized residuals; the latter have been corrected for differences in the SD of residuals depending on their position in the design. (Residuals corresponding to extreme *x*-values generally have a lower SD due to overfitting.) The third plot is of the square root of the absolute value of the standardized residuals; this reduces the skewness of the distribution and makes it much easier to detect if there might be a

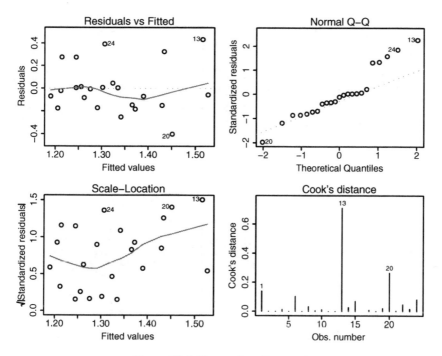

Figure 12.6. Regression diagnostics.

trend in the dispersion. The fourth plot is of "Cook's distance", which is a measure of the influence of each observation on the regression coefficients. We will return to Cook's distance shortly. Actually, this is not the default set of plots; the default replaces the fourth plot by a plot that contains the two components that enter into the calculation of Cook's distance, but this is harder to explain at this level.

The plots for the thuesen data show observation no. 13 as extreme in several respects. It has the largest residual as well as a prominent spike in the Cook's distance plot. Observation no. 20 also has a large residual, but not quite as conspicuous a Cook's distance.

```
> opar <- par(mfrow=c(2,2), mex=0.6, mar=c(4,4,3,2)+.3)
> plot(rstandard(lm.velo))
> plot(rstudent(lm.velo))
> plot(dffits(lm.velo),type="l")
> matplot(dfbetas(lm.velo),type="l", col="black")
> lines(sqrt(cooks.distance(lm.velo)), lwd=2)
> par(opar)
```

It is also possible to obtain individual diagnostics; a selection is shown in Figure 12.7. The function rstandard gives the standardized residuals

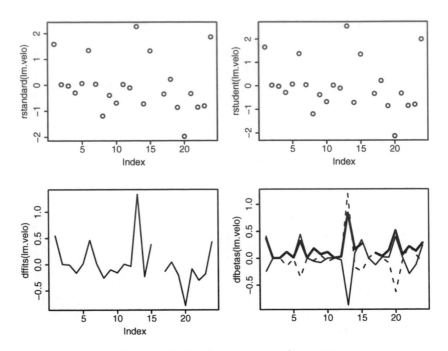

Figure 12.7. Further regression diagnostics.

discussed above. There is also rstudent, which gives *leave-out-one residuals*, in which the fitted value is calculated omitting the current point; if the model is correct, then these will follow a (Student's) *t* distribution. (Unfortunately, some texts use "studentized residuals" for residuals divided by their standard deviation; i.e., what rstandard calculates in R.) It takes a keen eye to see the difference between the two types of residuals, but the extreme residuals tend to be a little further out in the case of rstudent.

The function dffits expresses how much an observation affects the associated fitted value. As with the residuals, observations 13 and maybe 20 seem to stick out. Notice that there is a gap in the line. This is due to the missing observation 16 and the use of na.exclude. This looks a little awkward but has the advantage of making the *x*-axis match the observation number.

The function dfbetas gives the change in the estimated parameters if an observation is excluded relative to its standard error. It is a matrix, so matplot is useful to plot them all in one plot. Notice that observation 13 affects both α (the solid line) and β by nearly one standard error.

The name dfbetas refers to its use in multiple regression analysis, where you write the model as $y = \beta_0 + \beta_1 x_1 + \beta_2 x_2 + \cdots$. This gets a little confusing in a simple regression analysis, where the intercept is otherwise called α.

Cook's distance D calculated by `cooks.distance` is essentially a joint measure of the components of `dfbetas`. The exact procedure is to take the *unnormalized* change in coefficients and use the norm defined by the estimated covariance matrix for $\hat{\beta}$ and then divide by the number of coefficients. $\sqrt{D}$ is on the same scale as `dfbetas` and was added to that plot as a double-width line. (If you look inside the R functions for some of these quantities, you will find them apparently quite different from the descriptions above, but they are in fact the same, only computationally more efficient.)

Thus, the picture is that observation 13 seems to be influential. Let us look at the analysis without this observation.

We use the `subset` argument to `lm`, which, like other indexing operations, can be used with negative numbers to remove observations.

```
> summary(lm(short.velocity~blood.glucose, subset=-13))

Call:
lm(formula = short.velocity ~ blood.glucose, subset = -13)

Residuals:
     Min       1Q   Median       3Q      Max
-0.31346 -0.11136 -0.01247  0.06043  0.40794

Coefficients:
              Estimate Std. Error t value Pr(>|t|)
(Intercept)    1.18929    0.11061  10.752 9.22e-10 ***
blood.glucose  0.01082    0.01029   1.052    0.305
---
Signif. codes:  0 '***' 0.001 '**' 0.01 '*' 0.05 '.' 0.1 ' ' 1

Residual standard error: 0.193 on 20 degrees of freedom
  (1 observation deleted due to missingness)
Multiple R-squared: 0.05241,    Adjusted R-squared: 0.005026
F-statistic: 1.106 on 1 and 20 DF,  p-value: 0.3055
```

The relation practically vanished in thin air! The whole analysis actually hinges on a single observation. If the data and model are valid, then of course the original p-value is correct, and perhaps you could also say that there will always be influential observations in small data sets, but some caution in the interpretation does seem advisable.

The methods for finding influential observations and outliers are even more important in regression analysis with multiple descriptive variables. One of the big problems is how to present the quantities graphically in a sensible way. This might be done using three-dimensional plots (the add-

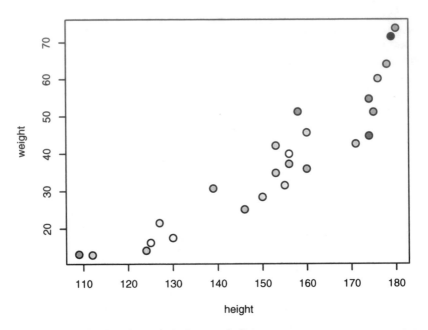

Figure 12.8. Cook's distance (colour coded) in `pemax ~ height + weight`.

on package `scatterplot3d` makes this possible), but you can get quite far using colour coding.

Here, we see how to display the value of Cook's distance (which is always positive) graphically for a model where `pemax` is described using `height` and `weight`, as in Figure 12.8:

```
> cookd <- cooks.distance(lm(pemax~height+weight))
> cookd <- cookd/max(cookd)
> cook.colors <- gray(1-sqrt(cookd))
> plot(height,weight,bg=cook.colors,pch=21,cex=1.5)
> points(height,weight,pch=1,cex=1.5)
```

The first line computes Cook's distance and the second scales it to a value between 0 and 1. Thereafter, a colour coding of the values in `cookd` is made with the function `gray`. The latter interprets its argument as the degree of whiteness, so if you want a large distance represented as black, you need to subtract the value from 1. Furthermore, it is convenient to take the square root of `cookd` because it is a quadratic distance measure (which in practice shows up in the form of too many white or nearly white points). Then a scatterplot of height versus weight is drawn with the cho-

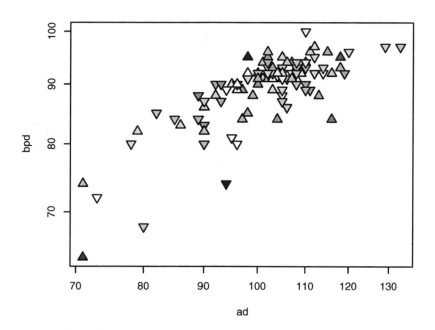

Figure 12.9. Studentized residuals in the Secher data, colour coded. Positive values are marked with upward-pointing triangles; negative ones point down.

sen colours. A filled plotting symbol in enlarged symbol size is used to get the grayscale to stand out more clearly.

You can use similar techniques to describe other influence measures. In the case of signed measures, you might use different symbols for positive and negative values. Here is an example on Studentized residuals in a data set describing birth weight as a function of abdominal and biparietal diameters determined by ultrasonography of the fetus immediately before birth, also used in Exercise 11.1 (Figure 12.9):

```
> attach(secher)
> rst <- rstudent(lm(log10(bwt)~log10(ad)+log10(bpd)))
> range(rst)
[1] -3.707509  3.674050
> rst <- rst/3.71
> plot(ad,bpd,log="xy",bg=gray(1-abs(rst)),
+       pch=ifelse(rst>0,24,25), cex=1.5)
```

12.9   Exercises

12.1   Set up an additive model for the `ashina` data (see Exercise 5.6) containing additive effects of subjects, period, and treatment. Compare the results with those obtained from t tests.

12.2   Perform a two-way analysis of variance on the `tb.dilute` data. Modify the model to have a dose effect that is linear in log dose. Compute a confidence interval for the slope. An alternative approach could be to calculate a slope for each animal and perform a test based on them. Compute a confidence interval for the mean slope, and compare it with the preceding result.

12.3   Consider the following definitions:

```
a <- gl(2, 2, 8)
b <- gl(2, 4, 8)
x <- 1:8
y <- c(1:4,8:5)
z <- rnorm(8)
```

Generate the model matrices for models `z ~ a*b`, `z ~ a:b`, etc. Discuss the implications. Carry out the model fits, and notice which models contain singularities.

12.4   (Advanced) In the `secretin` experiment, you may expect to find inter-individual differences not only in the level of glucose but also in the change induced by the injection of secretin. The factor `time.comb` combines time values at 30, 60, and 90 minutes. The factor `time20plus` combines all values from 20 minutes onward. Discuss the differences and relations among the following linear models:

```
attach(secretin)
model1 <- lm(gluc ~ person * time)
model2 <- lm(gluc ~ person + time)
model3 <- lm(gluc ~ person * time20plus + time)
model4 <- lm(gluc ~ person * time20plus + time.comb)
```

12.5   Analyze the blood pressure in the `bp.obese` data set as a function of obesity and gender.

12.6   Analyze the `vitcap2` data set using analysis of covariance. Revisit Exercise 5.2 and compare the conclusions. Try using the `drop1` function with `test="F"` instead of `summary` in this model.

12.7   In the `juul` data set make regression analyses for prepubescent children (Tanner stage 1) of $\sqrt{\text{igf1}}$ versus age separately for boys and girls. Compare the two regression lines.

12.8 Try `step` on the `kfm` data and discuss the result. One observation appears to be influential on the diagnostic plot for this model — explain why. What happens if you reduce the model further?

12.9 For the `juul` data, fit a model for `igf1` with interactions between age, sex, and Tanner stage for those under 25 years old. Explain the interpretation of this model. Hint: A plot of the fitted values against age should be helpful. Use diagnostic plots to evaluate possible transformations of the dependent variable: untransformed, log, or square root.

13

Logistic regression

Sometimes you wish to model *binary outcomes*, variables that can have only two possible values: diseased or nondiseased, and so forth. For instance, you want to describe the risk of getting a disease depending on various kinds of exposures. Chapter 8 discusses some simple techniques based on tabulation, but you might also want to model dose-response relationships (where the predictor is a continuous variable) or model the effect of multiple variables simultaneously. It would be very attractive to be able to use the same modelling techniques as for linear models.

However, it is not really attractive to use additive models for probabilities since they have a limited range and regression models could predict off-scale values below zero or above 1. It makes better sense to model the probabilities on a transformed scale; this is what is done in logistic regression analysis.

A linear model for transformed probabilities can be set up as

$$\text{logit } p = \beta_0 + \beta_1 x_1 + \beta_2 x_2 + \ldots \beta_k x_k$$

in which $\text{logit } p = \log[p/(1-p)]$ is the *log odds*. A constant additive effect on the logit scale corresponds to a constant odds ratio. The choice of the logit function is not the only one possible, but it has some mathematically convenient properties. Other choices do exist; the probit function (the quantile function of the normal distribution) or $\log(-\log p)$, which has a connection to survival analysis models.

P. Dalgaard, *Introductory Statistics with R*,
DOI: 10.1007/978-0-387-79054-1_13, © Springer Science+Business Media, LLC 2008

One thing to notice about the logistic model is that there is no error term as in linear models. We are modelling the probability of an event directly, and that in itself will determine the variability of the binary outcome. There is no variance parameter as in the normal distribution.

The parameters of the model can be estimated by the *method of maximum likelihood*. This is a quite general technique, similar to the least-squares method in that it finds a set of parameters that optimizes a goodness-of-fit criterion (in fact, the least-squares method itself is a slightly modified maximum-likelihood procedure). The *likelihood function* $L(\beta)$ is simply the probability of the entire observed data set for varying parameters.

The *deviance* is the difference between the maximized value of $-2 \log L$ and the similar quantity under a "maximal model" that fits data perfectly. Changes in deviance caused by a model reduction will be approximately χ^2-distributed with degrees of freedom equal to the change in the number of parameters.

In this chapter, we see how to perform logistic regression analysis in R. There naturally is quite a large overlap with the material on linear models since the description of models is quite similar, but there are also some special issues concerning deviance tables and the specification of models for pretabulated data.

13.1   Generalized linear models

Logistic regression analysis belongs to the class of *generalized linear models*. These models are characterized by their response distribution (here the binomial distribution) and a *link function*, which transfers the mean value to a scale in which the relation to background variables is described as linear ans additive. In a logistic regression analysis, the link function is logit $p = \log[p/(1-p)]$.

There are several other examples of generalized linear models; for instance, analysis of count data is often handled by the multiplicative Poisson model, where the link function is $\log \lambda$, with λ the mean of the Poisson-distributed observation. All of these models can be handled using the same algorithm, which also allows the user some freedom to define his or her own models by defining suitable link functions.

In R generalized linear models are handled by the `glm` function. This function is very similar to `lm`, which we have used many times for linear normal models. The two functions use essentially the same model formulas and extractor functions (`summary`, etc.), but `glm` also needs to have specified *which* generalized linear model is desired. This is done via the `family` argument. To specify a binomial model with logit link (i.e., logistic regression analysis), you write `family=binomial("logit")`.

13.2   Logistic regression on tabular data

In this section, we analyze the example concerning hypertension from Altman (1991, p. 353). First, we need to enter data, which is done as follows:

```
> no.yes <- c("No","Yes")
> smoking <- gl(2,1,8,no.yes)
> obesity <- gl(2,2,8,no.yes)
> snoring <- gl(2,4,8,no.yes)
> n.tot <- c(60,17,8,2,187,85,51,23)
> n.hyp <- c(5,2,1,0,35,13,15,8)
> data.frame(smoking,obesity,snoring,n.tot,n.hyp)
  smoking obesity snoring n.tot n.hyp
1      No      No      No    60     5
2     Yes      No      No    17     2
3      No     Yes      No     8     1
4     Yes     Yes      No     2     0
5      No      No     Yes   187    35
6     Yes      No     Yes    85    13
7      No     Yes     Yes    51    15
8     Yes     Yes     Yes    23     8
```

The `gl` function to "generate levels" was briefly introduced in Section 7.3. The first three arguments to `gl` are, respectively, the number of levels, the repeat count of each level, and the total length of the vector. A fourth argument can be used to specify the level names of the resulting factor. The result is apparent from the printout of the generated variables. They were put together in a data frame to get a nicer layout. Another way of generating a regular pattern like this is to use `expand.grid`:

```
> expand.grid(smoking=no.yes, obesity=no.yes, snoring=no.yes)
  smoking obesity snoring
1      No      No      No
2     Yes      No      No
3      No     Yes      No
4     Yes     Yes      No
5      No      No     Yes
6     Yes      No     Yes
7      No     Yes     Yes
8     Yes     Yes     Yes
```

R is able to fit logistic regression analyses for tabular data in two different ways. You have to specify the response as a matrix, where one column is the number of "diseased" and the other is the number of "healthy" (or "success" and "failure", depending on context).

```
> hyp.tbl <- cbind(n.hyp,n.tot-n.hyp)
> hyp.tbl
     n.hyp
[1,]     5   55
[2,]     2   15
[3,]     1    7
[4,]     0    2
[5,]    35  152
[6,]    13   72
[7,]    15   36
[8,]     8   15
```

The cbind function ("c" for "column") is used to bind variables together, columnwise, to form a matrix. Note that it would be a horrible mistake to use the total count for column 2 instead of the number of failures.

Then, you can specify the logistic regression model as

```
> glm(hyp.tbl~smoking+obesity+snoring,family=binomial("logit"))
```

Actually, "logit" is the default for binomial and the family argument is the second argument to glm, so it suffices to write

```
> glm(hyp.tbl~smoking+obesity+snoring,binomial)
```

The other way to specify a logistic regression model is to give the *proportion* of diseased in each cell:

```
> prop.hyp <- n.hyp/n.tot
> glm.hyp <- glm(prop.hyp~smoking+obesity+snoring,
+                  binomial,weights=n.tot)
```

It is necessary to give weights because R cannot see how many observations a proportion is based on.

As output, you get in either case (except for minor details)

```
Call:  glm(formula = hyp.tbl ~ smoking + obesity + snoring, ...

Coefficients:
(Intercept)    smokingYes    obesityYes    snoringYes
   -2.37766      -0.06777       0.69531       0.87194

Degrees of Freedom: 7 Total (i.e. Null);  4 Residual
Null Deviance:        14.13
Residual Deviance: 1.618          AIC: 34.54
```

which is in a minimal style similar to that used for printing lm objects. Also in the result of glm is some nonvisible information, which may be extracted with particular functions. You can, for instance, save the result of a fit of a generalized linear model in a variable and obtain a table of regression coefficients and so forth using summary:

```
> glm.hyp <- glm(hyp.tbl~smoking+obesity+snoring,binomial)
> summary(glm.hyp)

Call:
glm(formula = hyp.tbl ~ smoking + obesity + snoring, family ...

Deviance Residuals:
       1         2         3         4         5         6
-0.04344   0.54145  -0.25476  -0.80051   0.19759  -0.46602
       7         8
-0.21262   0.56231

Coefficients:
             Estimate Std. Error z value Pr(>|z|)
(Intercept) -2.37766    0.38018  -6.254   4e-10 ***
smokingYes  -0.06777    0.27812  -0.244   0.8075
obesityYes   0.69531    0.28509   2.439   0.0147 *
snoringYes   0.87194    0.39757   2.193   0.0283 *
---
Signif. codes:  0 '***' 0.001 '**' 0.01 '*' 0.05 '.' 0.1 ' ' 1

(Dispersion parameter for binomial family taken to be 1)

    Null deviance: 14.1259  on 7  degrees of freedom
Residual deviance:  1.6184  on 4  degrees of freedom
AIC: 34.537

Number of Fisher Scoring iterations: 4
```

In the following, we go through the components of summary output for generalized linear models:

```
Call:
glm(formula = hyp.tbl ~ smoking + obesity + snoring, family = ...
```

As usual, we start off with a repeat of the model specification. Obviously, more interesting is when the output is not viewed in connection with the function call that generated it.

```
Deviance Residuals:
       1         2         3         4         5         6
-0.04344   0.54145  -0.25476  -0.80051   0.19759  -0.46602
       7         8
-0.21262   0.56231
```

This is the contribution of each cell of the table to the deviance of the model (the deviance corresponds to the sum of squares in linear normal models), with a sign according to whether the observation is larger or smaller than expected. They can be used to pinpoint cells that are particularly poorly fitted, but you have to be wary of the interpretation in sparse tables.

```
Coefficients:
             Estimate Std. Error z value Pr(>|z|)
(Intercept) -2.37766    0.38018  -6.254    4e-10 ***
smokingYes  -0.06777    0.27812  -0.244   0.8075
obesityYes   0.69531    0.28509   2.439   0.0147 *
snoringYes   0.87194    0.39757   2.193   0.0283 *
---
Signif. codes:  0 `***' 0.001 `**' 0.01 `*' 0.05 `.' 0.1 ` ' 1

(Dispersion parameter for binomial family taken to be 1)
```

This is the table of primary interest. Here, we get estimates of the regression coefficients, standard errors of same, and tests for whether each regression coefficient can be assumed to be zero. The layout is nearly identical to the corresponding part of the lm output.

The note about the dispersion parameter is related to the fact that the binomial variance depends entirely on the mean. There is no scale parameter like the variance in the normal distribution.

```
    Null deviance: 14.1259  on 7  degrees of freedom
Residual deviance:  1.6184  on 4  degrees of freedom
AIC: 34.537
```

"Residual deviance" corresponds to the residual sum of squares in ordinary regression analyses which is used to estimate the standard deviation about the regression line. In binomial models, however, the standard deviation of the observations is known, and you can therefore use the deviance in a test for model specification. The AIC (Akaike information criterion) is a measure of goodness of fit that takes the number of fitted parameters into account.

R is reluctant to associate a p-value with the deviance. This is just as well because no exact p-value can be found, only an approximation that is valid for large expected counts. In the present case, there are actually a couple of places where the expected cell count is rather small.

The asymptotic distribution of the residual deviance is a χ^2 distribution with the stated degrees of freedom, so even though the approximation may be poor, nothing in the data indicates that the model is wrong (the 5% significance limit is at 9.49 and the value found here is 1.62).

The null deviance is the deviance of a model that contains only the intercept (that is, describes a fixed probability, here for hypertension, in all cells). What you would normally be interested in is the difference from the residual deviance, here $14.13 - 1.62 = 12.51$, which can be used for a joint test for whether any effects are present in the model. In the present case, a p-value of approximately 0.6% is obtained.

```
Number of Fisher Scoring iterations: 4
```

This refers to the actual fitting procedure and is a purely technical item. There is no statistical information in it, but you should keep an eye on whether the number of iterations becomes too large because that might be a sign that the model is too complex to fit based on the available data. Normally, glm halts the fitting procedure if the number of iterations exceeds 25, but it is possible to configure the limit.

The fitting procedure is *iterative* in that there is no explicit formula that can be used to compute the estimates, only a set of equations that they should satisfy. However, there is an approximate solution of the equations if you supply an initial guess at the solution. This solution is then used as a starting point for an improved solution, and the procedure is repeated until the guesses are sufficiently stable.

A table of correlations between parameter estimates can be obtained via the optional argument corr=T to summary (this also works for linear models). It looks like this:

```
Correlation of Coefficients:
           (Intercept) smokingYes obesityYes
smokingYes    -0.1520
obesityYes    -0.1361 -9.499e-05
snoringYes    -0.8965 -6.707e-02    -0.07186
```

It is seen that the correlation between the estimates is fairly small, so that it may be expected that removing a variable from the model does not change the coefficients and p-values for other variables much. (The correlations between the regression coefficients and intercept are not very informative; they mostly relate to whether the variable in question has many or few observations in the "Yes" category.)

The z test in the table of regression coefficients immediately shows that the model can be simplified by removing smoking. The result then looks as follows (abbreviated):

```
> glm.hyp <- glm(hyp.tbl~obesity+snoring,binomial)
> summary(glm.hyp)

...
```

```
Coefficients:
              Estimate Std. Error z value Pr(>|z|)
(Intercept)   -2.3921      0.3757  -6.366 1.94e-10 ***
obesityYes     0.6954      0.2851   2.440   0.0147 *
snoringYes     0.8655      0.3967   2.182   0.0291 *
```

13.2.1   The analysis of deviance table

Deviance tables correspond to ANOVA tables for multiple regression analyses and are generated like these with the anova function:

```
> glm.hyp <- glm(hyp.tbl~smoking+obesity+snoring,binomial)
> anova(glm.hyp, test="Chisq")
Analysis of Deviance Table

Model: binomial, link: logit

Response: hyp.tbl

Terms added sequentially (first to last)


        Df Deviance Resid. Df Resid. Dev P(>|Chi|)
NULL                       7    14.1259
smoking  1   0.0022        6    14.1237    0.9627
obesity  1   6.8274        5     7.2963    0.0090
snoring  1   5.6779        4     1.6184    0.0172
```

Notice that the Deviance column gives *differences* between models as variables are added to the model in turn. The deviances are approximately χ^2-distributed with the stated degrees of freedom. It is necessary to add the test="chisq" argument to get the approximate χ^2 tests.

Since the snoring variable on the last line is significant, it may not be removed from the model and we cannot use the table to justify model reductions. If, however, the terms are rearranged so that smoking comes last, we get a deviance-based test for removal of that variable:

```
> glm.hyp <- glm(hyp.tbl~snoring+obesity+smoking,binomial)
> anova(glm.hyp, test="Chisq")
...
        Df Deviance Resid. Df Resid. Dev P(>|Chi|)
NULL                       7    14.1259
snoring  1   6.7887        6     7.3372    0.0092
obesity  1   5.6591        5     1.6781    0.0174
smoking  1   0.0597        4     1.6184    0.8069
```

From this you can read that smoking is removable, whereas obesity is not, after removal of smoking.

For good measure, you should also set up the analysis with the two remaining explanatory variables interchanged, so that you get a test of
whether snoring may be removed from a model that also contains
obesity:

```
> glm.hyp <- glm(hyp.tbl~obesity+snoring,binomial)
> anova(glm.hyp, test="Chisq")
...
         Df Deviance Resid. Df Resid. Dev P(>|Chi|)
NULL                       7     14.1259
obesity  1    6.8260        6      7.2999    0.0090
snoring  1    5.6218        5      1.6781    0.0177
```

An alternative method is to use drop1 to try removing one term at a time:

```
> drop1(glm.hyp, test="Chisq")
Single term deletions

Model:
hyp.tbl ~ obesity + snoring
         Df Deviance     AIC    LRT Pr(Chi)
<none>         1.678  32.597
obesity  1     7.337  36.256  5.659 0.01737 *
snoring  1     7.300  36.219  5.622 0.01774 *
---
Signif. codes:  0 '***' 0.001 '**' 0.01 '*' 0.05 '.' 0.1 ' ' 1
```

Here LRT is the likelihood ratio test, another name for the deviance
change.

The information in the deviance tables is fundamentally the same as that
given by the z tests in the table of regression coefficients. The results may
differ due to the use of different approximations, though. From theoretical
considerations, the deviance test is preferable, but in practice the difference is often small because of the large-sample approximation $\chi^2 \approx z^2$
for tests with a single degree of freedom. However, to test factors with
more than two categories, you have to use the deviance table because the
z tests only relate to some of the possible group comparisons. Also, the
small-sample situation requires special attention; see the next section.

13.2.2 Connection to test for trend

In Chapter 8, we considered tests for comparing relative frequencies using
prop.test and prop.trend.test, in particular the example of caesarean section versus shoe size. This example can also be analyzed as a
logistic regression analysis on a "shoe score", which — for want of a better idea — may be chosen as the group number. This gives essentially the
same analysis in the sense that the same models are involved.

```
> caesar.shoe
    <4   4 4.5   5 5.5   6+
Yes  5   7    6   7   8   10
No  17  28   36  41   46  140
> shoe.score <- 1:6
> shoe.score
[1] 1 2 3 4 5 6

> summary(glm(t(caesar.shoe)~shoe.score,binomial))
...
Coefficients:
            Estimate Std. Error z value Pr(>|z|)
(Intercept) -0.87058    0.40506  -2.149  0.03161 *
shoe.score  -0.25971    0.09361  -2.774  0.00553 **
---
Signif. codes:  0 '***' 0.001 '**' 0.01 '*' 0.05 '.' 0.1 ' ' 1

(Dispersion parameter for binomial family taken to be 1)

    Null deviance: 9.3442  on 5  degrees of freedom
Residual deviance: 1.7845  on 4  degrees of freedom
AIC: 27.616
...
```

Notice that `caesar.shoe` had to be transposed with `t(...)`, so that the matrix was "stood on its end" in order to be used as the response variable by `glm`.

You can also write the results in a deviance table

```
> anova(glm(t(caesar.shoe)~shoe.score,binomial))
...
            Df Deviance Resid. Df Resid. Dev
NULL                          5      9.3442
shoe.score   1   7.5597        4      1.7845
```

from the last line of which you see that there is no significant deviation from linearity (1.78 on 4 degrees of freedom), whereas `shoe.score` has a significant contribution.

For comparison, the previous analyses using standard tests are repeated:

```
> caesar.shoe.yes <- caesar.shoe["Yes",]
> caesar.shoe.no <- caesar.shoe["No",]
> caesar.shoe.total <- caesar.shoe.yes+caesar.shoe.no
> prop.trend.test(caesar.shoe.yes,caesar.shoe.total)
        Chi-squared Test for Trend in Proportions
...
X-squared = 8.0237, df = 1, p-value = 0.004617

> prop.test(caesar.shoe.yes,caesar.shoe.total)
```

```
    6-sample test for equality of proportions without
    continuity correction
...
X-squared = 9.2874, df = 5, p-value = 0.09814
...
Warning message:
In prop.test(caesar.shoe.yes, caesar.shoe.total) :
  Chi-squared approximation may be incorrect
```

The 9.29 from `prop.test` corresponds to the 9.34 in residual deviance from a `NULL` model, whereas the 8.02 in the trend test corresponds to the 7.56 in the test of significance of `shoe.score`. Thus, the tests do not give exactly the same result but generally *almost* the same. Theoretical considerations indicate that the specialized trend test is probably slightly better than the regression-based test. However, testing the linearity by subtracting the two χ^2 tests is definitely not as good as the real test for linearity.

13.3   Likelihood profiling

The z tests in the summary output are based on the *Wald approximation*, which calculates what the approximate standard error of the parameter estimate would be if the true values of the parameters were equal to the estimates. In large data sets, this is fine because the result is nearly the same for all parameter values that fit the data reasonably well. In smaller data sets, however, the difference between the Wald tests and the likelihood ratio test can be considerable.

This also affects the calculation of confidence intervals since these are based on inverting the tests, giving a set of parameter values that are not rejected by a statistical test. As an alternative to the Wald-based $\pm 1.96 \times$ s.e. technique, the `MASS` package allows you to compute intervals that are based on inverting the likelihood ratio test. In practice, this works like this

```
> confint(glm.hyp)
Waiting for profiling to be done...
                2.5 %     97.5 %
(Intercept) -3.2102369 -1.718143
obesityYes    0.1254382  1.246788
snoringYes    0.1410865  1.715860
```

The standard type of result can be obtained using `confint.default`. The difference in this case is not very large, although visible in the lines relating to snoring and the intercept:

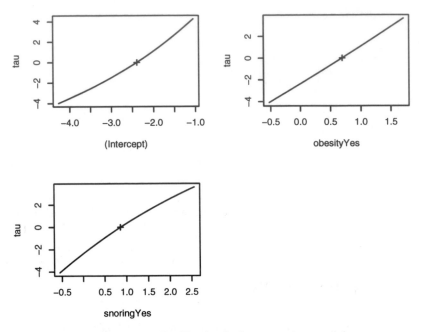

Figure 13.1. Profile plot for hypertension model.

```
> confint.default(glm.hyp)
                  2.5 %      97.5 %
(Intercept) -3.12852108  -1.655631
obesityYes   0.13670388   1.254134
snoringYes   0.08801498   1.642902
```

The way this works is via *likelihood profiling*. For a set of trial values of the parameter, the likelihood is maximized over the other parameters in the model. The result can be displayed in a profile plot as follows:

```
> library(MASS)
> plot(profile(glm.hyp))
```

Notice that we need to load the MASS package at this point. (The function was used by confint earlier on, but without putting it on the search path.)

The plots require a little explanation. The quantity on the y-axis, labelled tau, is the signed square root of the likelihood ratio test.

$$\tau(\beta) = \text{sgn}(\beta - \hat{\beta})\sqrt{-2(\ell(\beta) - \ell(\hat{\beta}))}$$

Here ℓ denotes the profile log-likelihood. The main idea is that when the profile likelihood function is approximately quadratic, $\tau(\beta)$ is approximately linear. Conversely, likelihood functions not well approximated by a quadratic show up as nonlinear profile plots.

One important thing to notice, though, is that although the profiling method will capture nonquadratic behaviour of the likelihood function, confidence intervals based on the likelihood ratio test will always be limited in accuracy by the approximation of the distribution of the test.

13.4   Presentation as odds-ratio estimates

In parts of the epidemiological literature, it has become traditional to present logistic regression analyses in terms of odds ratios. In the case of a quantitative covariate, this means odds ratio per unit change in the covariate. That is, the antilogarithm (exp) of the regression coefficients is given instead of the coefficients themselves. Since standard errors make little sense after the transformation, it is also customary to give confidence intervals instead. This can be obtained quite easily as follows:

```
> exp(cbind(OR=coef(glm.hyp), confint(glm.hyp)))
Waiting for profiling to be done...
                   OR       2.5 %      97.5 %
(Intercept) 0.09143963 0.04034706 0.1793989
obesityYes  2.00454846 1.13364514 3.4791490
snoringYes  2.37609483 1.15152424 5.5614585
```

The (Intercept) is really the odds of hypertension (for the not snoring non-obese) and not an odds ratio.

13.5   Logistic regression using raw data

In this section, we again use Anders Juul's data (see p. 85). For easy reference, here is how to read data and convert the variables that describe groupings into factors (this time slightly simplified):

```
> juul$menarche <- factor(juul$menarche, labels=c("No","Yes"))
> juul$tanner <- factor(juul$tanner)
```

In the following, we look at menarche as the response variable. This variable indicates for each girl whether or not she has had her first period. It is coded 1 for "no" and 2 for "yes". It is convenient to look at a subset of data consisting of 8–20-year-old girls. This can be extracted as follows:

```
> juul.girl <- subset(juul,age>8 & age<20 &
+                          complete.cases(menarche))
> attach(juul.girl)
```

For obvious reasons, no boys have a nonmissing menarche, so it is not necessary to select on gender explicitly.

Then you can analyze menarche as a function of age like this:

```
> summary(glm(menarche~age,binomial))
Call:
glm(formula = menarche ~ age, family = binomial)

Deviance Residuals:
    Min        1Q     Median        3Q       Max
-2.32759   -0.18998   0.01253    0.12132   2.45922

Coefficients:
             Estimate Std. Error z value Pr(>|z|)
(Intercept) -20.0132     2.0284  -9.867   <2e-16 ***
age           1.5173     0.1544   9.829   <2e-16 ***
---
Signif. codes:  0 '***' 0.001 '**' 0.01 '*' 0.05 '.' 0.1 ' ' 1

(Dispersion parameter for binomial family taken to be 1)

    Null deviance: 719.39  on 518  degrees of freedom
Residual deviance: 200.66  on 517  degrees of freedom
AIC: 204.66

Number of Fisher Scoring iterations: 7
```

The response variable menarche is a factor with two levels, where the last level is considered the event. It also works to use a variable that has the values 0 and 1 (but *not*, for instance, 1 and 2!).

Notice that from this model you can estimate the median menarcheal age as the age where logit $p = 0$. A little thought (solve $-20.0132 + 1.5173 \times$ age $= 0$) reveals that it is $20.0132/1.5173 = 13.19$ years.

You should not pay too much attention to the deviance residuals in this case since they automatically become large in every case where the fitted probability "goes against" the observations (which is bound to happen in some cases). The residual deviance is also difficult to interpret when there is only one observation per cell.

A hint of a more complicated analysis is obtained by including the Tanner stage of puberty in the model. You should be warned that the exact interpretation of such an analysis is quite tricky and *qualitatively* different from the analysis of menarche as a function of age. It can be used for prediction purposes (although asking the girl whether she has had her first period would likely be much easier than determining her Tanner stage!), but the interpretation of the terms is not clear-cut.

```
> summary(glm(menarche~age+tanner,binomial))
...
Coefficients:
            Estimate Std. Error z value Pr(>|z|)
(Intercept) -13.7758     2.7630  -4.986 6.17e-07 ***
age           0.8603     0.2311   3.723 0.000197 ***
tanner2      -0.5211     1.4846  -0.351 0.725609
tanner3       0.8264     1.2377   0.668 0.504313
tanner4       2.5645     1.2172   2.107 0.035132 *
tanner5       5.1897     1.4140   3.670 0.000242 ***
...
```

Notice that there is no joint test for the effect of tanner. There are a couple of significant z-values, so you would expect that the tanner variable has some effect (which, of course, you would probably expect even in the absence of data!). The formal test, however, must be obtained from the deviances:

```
> drop1(glm(menarche~age+tanner,binomial),test="Chisq")
...
        Df Deviance     AIC     LRT   Pr(Chi)
<none>      106.599 118.599
age      1  124.500 134.500  17.901 2.327e-05 ***
tanner   4  161.881 165.881  55.282 2.835e-11 ***
...
```

Clearly, both terms are highly significant.

13.6   Prediction

The predict function works for generalized linear models, too. Let us first consider the hypertension example, where data were given in tabular form:

```
> predict(glm.hyp)
         1          2          3          4          5          6
-2.3920763 -2.3920763 -1.6966575 -1.6966575 -1.5266180 -1.5266180
         7          8
-0.8311991 -0.8311991
```

Recall that smoking was eliminated from the model, which is why the expected values come in identical pairs.

These numbers are on the logit scale, which reveals the additive structure. Notice that $2.392 - 1.697 = 1.527 - 0.831 = 0.695$ (except for roundoff error), which is exactly the regression coefficient to obesity. Likewise, the regression coefficient to snoring is obtained by looking at the differences $2.392 - 1.527 = 1.697 - 0.831 = 0.866$.

To get predicted values on the response scale (i.e., probabilities), use the type="response" argument to predict:

```
> predict(glm.hyp, type="response")
         1          2          3          4          5          6
0.08377892 0.08377892 0.15490233 0.15490233 0.17848906 0.17848906
         7          8
0.30339158 0.30339158
```

These may also be obtained using fitted, although you then cannot use the techniques for predicting on new data, etc.

In the analysis of menarche, the primary interest is probably in seeing a plot of the expected probabilities versus age (Figure 13.2). A crude plot could be obtained using something like

```
plot(age, fitted(glm(menarche~age,binomial)))
```

(it will look better if a different plotting symbol in a smaller size, using the pch and cex arguments, is used) but here is a more ambitious plan:

```
> glm.menarche <- glm(menarche~age, binomial)
> Age <- seq(8,20,.1)
> newages <- data.frame(age=Age)
> predicted.probability <- predict(glm.menarche,
+                                  newages,type="resp")
> plot(predicted.probability ~ Age, type="l")
```

This is Figure 13.2. Recall that seq generates equispaced vectors, here ages from 8 to 20 in steps of 0.1, so that connecting the points with lines will give a nearly smooth curve.

13.7   Model checking

For tabular data it is obvious to try to compare observed and fitted proportions. In the hypertension example you get

```
> fitted(glm.hyp)
         1          2          3          4          5          6
0.08377892 0.08377892 0.15490233 0.15490233 0.17848906 0.17848906
         7          8
0.30339158 0.30339158
> prop.hyp
```

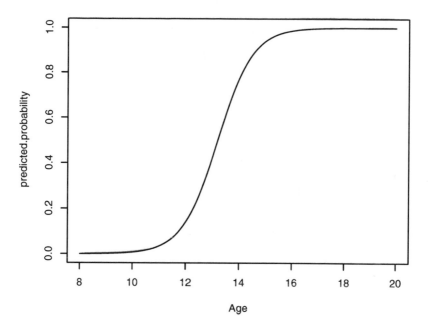

Figure 13.2. Fitted probability of menarche having occurred.

```
[1] 0.08333333 0.11764706 0.12500000 0.00000000 0.18716578
[6] 0.15294118 0.29411765 0.34782609
```

The problem with this is that you get no feeling for how well the relative frequencies are determined. It can be better to look at observed and expected *counts* instead. The former can be computed as

```
> fitted(glm.hyp)*n.tot
         1           2           3           4           5           6
 5.0267351   1.4242416   1.2392186   0.3098047  33.3774535  15.1715698
         7           8
15.4729705   6.9780063
```

and to get a nice print for the comparison, you can use

```
> data.frame(fit=fitted(glm.hyp)*n.tot,n.hyp,n.tot)
         fit n.hyp n.tot
1  5.0267351     5    60
2  1.4242416     2    17
3  1.2392186     1     8
4  0.3098047     0     2
5 33.3774535    35   187
6 15.1715698    13    85
```

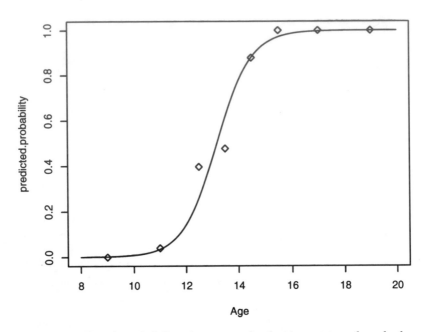

Figure 13.3. Fitted probability for menarche having occurred and observed proportion in age groups.

```
7 15.4729705    15    51
8  6.9780063     8    23
```

Notice that the discrepancy in cell 4 between 15% expected and 0% observed really is that there are 0 hypertensives out of 2 in a cell where the model yields an expectation of 0.3 hypertensives!

For complex models with continuous background variables, it becomes more difficult to perform an adequate model check. It is especially a hindrance that nothing really corresponds to a residual plot when the observations have only two different values.

Consider the example of the probability of menarche as a function of age. The problem here is whether the relation can really be assumed linear on the logit scale. For this case, you might try subdividing the x-axis in a number of intervals and see how the counts in each interval fit with the expected probabilities. This is presented graphically in Figure 13.3. Notice that the code *adds* points to Figure 13.2, which you are assumed not to have deleted at this point.

```
> age.group <- cut(age,c(8,10,12,13,14,15,16,18,20))
```

```
> tb <- table(age.group,menarche)
> tb
         menarche
age.group  No Yes
  (8,10]  100   0
  (10,12]  97   4
  (12,13]  32  21
  (13,14]  22  20
  (14,15]   5  36
  (15,16]   0  31
  (16,18]   0 105
  (18,20]   0  46
> rel.freq <- prop.table(tb,1)[,2]
> rel.freq
     (8,10]    (10,12]    (12,13]    (13,14]    (14,15]    (15,16]
0.00000000 0.03960396 0.39622642 0.47619048 0.87804878 1.00000000
    (16,18]    (18,20]
1.00000000 1.00000000
> points(rel.freq ~ c(9,11,12.5,13.5,14.5,15.5,17,19),pch=5)
```

The technique used above probably requires some explanation. First, cut is used to define the factor age.group, which describes a grouping into age intervals. Then a crosstable tb is formed from menarche and age.group. Using prop.table, the numbers are expressed relative to the row total, and column 2 of the resulting table is extracted. This contains the relative proportion in each age group of girls for whom menarche has occurred. Finally, a plot of expected probabilities is made, overlaid by the observed proportions.

The plot looks reasonable on the whole, although the observed proportion among 12–13-year-olds appears a bit high and the proportion among 13–14-year-olds is a bit too low.

But how do you evaluate whether the deviation is larger than what can be expected from the statistical variation? One thing to try is to extend the model with a factor that describes a division into intervals. It is not practical to use the full division of age.group because there are cells where either none or all of the girls have had their menarche.

We therefore try a division into four groups, with cutpoints at 12, 13, and 14 years, and add this factor to the model containing a linear age effect.

```
> age.gr <- cut(age,c(8,12,13,14,20))
> summary(glm(menarche~age+age.gr,binomial))
...
Coefficients:
               Estimate Std. Error z value Pr(>|z|)
(Intercept)    -21.5683     5.0645  -4.259 2.06e-05 ***
age              1.6250     0.4416   3.680 0.000233 ***
age.gr(12,13]    0.7296     0.7856   0.929 0.353024
age.gr(13,14]   -0.5219     1.1184  -0.467 0.640765
```

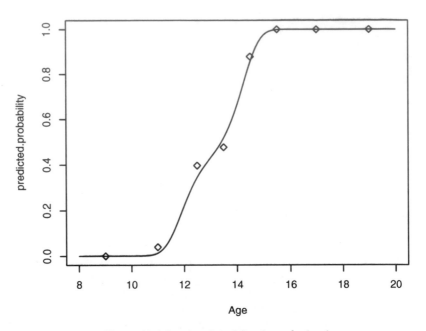

Figure 13.4. Logit-cubical fit of menarche data.

```
age.gr(14,20]    0.2751      1.6065    0.171 0.864053
...

> anova(glm(menarche~age+age.gr,binomial))
...
        Df Deviance Resid. Df Resid. Dev
NULL                    518       719.39
age      1   518.73    517       200.66
age.gr   3     8.06    514       192.61
> 1-pchisq(8.058,3)
[1] 0.04482811
```

That is, the addition of the grouping actually does give a significantly better deviance. The effect is not highly significant, but since the deviation concerns the ages where "much happens", you should probably be cautious about postulating a logit-linear age effect.

Another possibility is to try a polynomial regression model. Here you need at least a third-degree polynomial to describe the apparent stagnation of the curve around 13 years of age. We do not look at this in great detail, but just show part of the output and in Figure 13.4 a graphical presentation of the model.

```
> anova(glm(menarche~age+I(age^2)+I(age^3)+age.gr,binomial))
...
            Df Deviance Resid. Df Resid. Dev
NULL                        518       719.39
age          1   518.73     517       200.66
I(age^2)     1     0.05     516       200.61
I(age^3)     1     8.82     515       191.80
age.gr       3     3.34     512       188.46
Warning messages:
1: In glm.fit(x = X, y = Y, weights = weights, .... :
   fitted probabilities numerically 0 or 1 occurred
2: In method(x = x[, varseq <= i, drop = FALSE], .... :
   fitted probabilities numerically 0 or 1 occurred
> glm.menarche <- glm(menarche~age+I(age^2)+I(age^3), binomial)
Warning message:
In glm.fit(x = X, y = Y, weights = weights, start = start, .... :
   fitted probabilities numerically 0 or 1 occurred
> predicted.probability <-
+      predict(glm.menarche, newages, type="resp")
> plot(predicted.probability ~ Age, type="l")
> points(rel.freq~c(9,11,12.5,13.5,14.5,15.5,17,19), pch=5)
```

The warnings about fitted probabilities of 0 or 1 occur because the cubic term makes the logit tend much faster to $\pm\infty$ than the linear model did. There are two occurrences for the anova call because two of the models include the cubic term.

The thing to note in the deviance table is that the cubic term gives a substantial improvement of the deviance, but once that is included, the age grouping gives no additional improvement. The plot should speak for itself.

13.8 Exercises

13.1 In the malaria data set, analyze the risk of malaria with age and log-transformed antibody level as explanatory variables.

13.2 Fit a logistic regression model to the graft.vs.host data set, predicting the gvhd response. Use different transformations of the index variable. Reduce the model using backwards elimination.

13.3 In the analyses of the malaria and graft.vs.host data, try using the confint function to find improved confidence intervals for the regression coefficients.

13.4 Following up on Exercise 8.2 about "Rocky Mountain spotted fever", splitting the data by age groups gives the table below. Does this confirm the earlier analysis?

Age Group	Western Type Total	Western Type Fatal	Eastern Type Total	Eastern Type Fatal
Under 15	108	13	310	40
15–39	264	40	189	21
40 or above	375	157	162	61
	747	210	661	122

13.5 A *probit* regression is just like a logistic regression but uses a different link function. Try the analysis of the `menarche` variable in the `juul` data set with this link. Does the fit improve?

14

Survival analysis

The analysis of lifetimes is an important topic within biology and medicine in particular but also in reliability analysis with engineering applications. Such data are often highly nonnormally distributed, so that the use of standard linear models is problematic.

Lifetime data are often *censored:* You do not know the exact lifetime, only that it is longer than a given value. For instance, in a cancer trial, some people are lost to follow-up or simply live beyond the study period. It is an error to ignore the censoring in the statistical analysis, sometimes with extreme consequences. Consider, for instance, the case where a new treatment is introduced towards the end of the study period, so that nearly all the observed lifetimes will be cut short.

14.1   Essential concepts

Let X be the true lifetime and T a censoring time. What you observe is the minimum of X and T together with an indication of whether it is one or the other. T can be a random variable or a fixed time depending on context, but if it is random, then it should generally be *noninformative* for the methods we describe here to be applicable. Sometimes "dead from other causes" is considered a censoring event for the mortality of a given

P. Dalgaard, *Introductory Statistics with R*,
DOI: 10.1007/978-0-387-79054-1_14, © Springer Science+Business Media, LLC 2008

disease, and in those cases it is particularly important to ensure that these other causes are unassociated with the disease state.

The *survival function* $S(t)$ measures the probability of being alive at a given time. It is really just 1 minus the cumulative distribution function for X, $1 - F(t)$.

The *hazard function* or *force of mortality* $h(t)$ measures the (infinitesimal) risk of dying within a short interval of time t, given that the subject is alive at time t. If the lifetime distribution has density f, then $h(t) = f(t)/S(t)$. This is often considered a more fundamental quantity than (say) the mean or median of the survival distribution and is used as a basis for modelling.

14.2   Survival objects

We use the package `survival`, written by Terry Therneau and ported to R by Thomas Lumley. The package implements a large number of advanced techniques. For the present purposes, we use only a small subset of it.

To load `survival`, use

```
> library(survival)
```

(This may produce a harmless warning about masking the `lung` data set from the `ISwR` package.)

The routines in `survival` work with objects of class `"Surv"`, which is a data structure that combines times and censoring information. Such objects are constructed using the `Surv` function, which takes two arguments: an observation time and an event indicator. The latter can be coded as a logical variable, a 0/1 variable, or a 1/2 variable. The latter coding is not recommended since `Surv` will assume 0/1 coding if all values are 1.

Actually, `Surv` can also be used with three arguments for dealing with data that have a start time as well as an end time ("staggered entry") and also interval censored data (where you know that an event happened between two dates, as happens, for instance, in repeated testing for a disease) can be handled.

We use the data set `melanom` collected by K. T. Drzewiecki and reproduced in Andersen et al. (1991). The data become accessible as follows:

```
> attach(melanom)
> names(melanom)
```

```
[1] "no"      "status" "days"    "ulc"     "thick"  "sex"
```

The variable `status` is an indicator of the patient's status by the end of the study: 1 means "dead from malignant melanoma", 2 means "alive on January 1, 1978", and 3 means "dead from other causes". The variable `days` is the observation time in days, `ulc` indicates (1 for present and 2 for absent) whether the tumor was ulcerated, `thick` is the thickness in 1/100 mm, and `sex` contains the gender of the patient (1 for women and 2 for men).

We want to create a `Surv` object in which we consider the values 2 and 3 of the `status` variable as censorings. This is done as follows:

```
> Surv(days, status==1)
  [1]   10+    30+    35+    99+   185    204    210    232    232+   279
 [11]  295    355+   386    426   469    493+   529    621    629    659
 [21]  667    718    752    779   793    817    826+   833    858    869
...
[181] 3476+ 3523+ 3667+ 3695+ 3695+ 3776+ 3776+ 3830+ 3856+ 3872+
[191] 3909+ 3968+ 4001+ 4103+ 4119+ 4124+ 4207+ 4310+ 4390+ 4479+
[201] 4492+ 4668+ 4688+ 4926+ 5565+
```

Associated with the `Surv` objects is a print method that displays the objects in the format above, with a '+' marking censored observations. For example, `10+` means that the patient did not die from melanoma within 10 days and was then unavailable for further study (in fact, he died from other causes), whereas `185` means that the patient died from the disease a little over half a year after his operation.

Notice that the second argument to `Surv` is a logical vector; `status==1` is TRUE for those who died of malignant melanoma and FALSE otherwise.

14.3   Kaplan–Meier estimates

The Kaplan–Meier estimator allows the computation of an estimated survival function in the presence of right-censoring. It is also called the *product-limit estimator* because one way of describing the procedure is that it multiplies together conditional survival curves for intervals in which there are either no censored observations or no deaths. This becomes a step function where the estimated survival is reduced by a factor $(1 - 1/R_t)$ if there is a death at time t and a population of R_t is still alive and uncensored at that time.

Computing the Kaplan–Meier estimator for the survival function is done with a function called `survfit`. In its simplest form, it takes just a single argument, namely a `Surv` object. It returns a `survfit` object. As described above, we consider "dead from other causes" a kind of censoring and do as follows:

```
> survfit(Surv(days,status==1))
Call: survfit(formula = Surv(days, status == 1))

      n   events   median 0.95LCL 0.95UCL
    205       57      Inf     Inf     Inf
```

As is seen, using `survfit` by itself is not very informative (just as the printed output of a "bare" `lm` is not). You get a couple of summary statistics and an estimate of the median survival, and in this case the latter is not even interesting because the estimate is infinite. The survival curve does not cross the 50% mark before all patients are censored.

To see the actual Kaplan–Meier estimate, use `summary` on the `survfit` object. We first save the `survfit` object into a variable, here named `surv.all` because it contains the raw survival function for all patients without regard to patient characteristics.

```
> surv.all <- survfit(Surv(days,status==1))
> summary(surv.all)
Call: survfit(formula = Surv(days, status == 1))

time n.risk n.event survival std.err lower 95% CI upper 95% CI
 185    201       1    0.995 0.00496        0.985        1.000
 204    200       1    0.990 0.00700        0.976        1.000
 210    199       1    0.985 0.00855        0.968        1.000
 232    198       1    0.980 0.00985        0.961        1.000
 279    196       1    0.975 0.01100        0.954        0.997
 295    195       1    0.970 0.01202        0.947        0.994
...
2565     63       1    0.689 0.03729        0.620        0.766
2782     57       1    0.677 0.03854        0.605        0.757
3042     52       1    0.664 0.03994        0.590        0.747
3338     35       1    0.645 0.04307        0.566        0.735
```

This contains the values of the survival function at the event times. The censoring times are not displayed but are contained in the `survfit` object and can be obtained by passing `censored=T` to `summary` (see the help page for `summary.survfit` for such details).

The Kaplan–Meier estimate is the step function whose jump points are given in `time` and whose values right after a jump are given in `survival`. Additionally, both an estimate of the standard error of the curve and a (pointwise) confidence interval for the true curve are given.

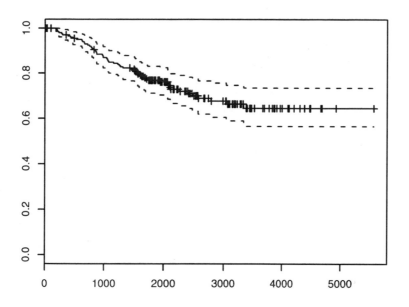

Figure 14.1. Kaplan–Meier plot for melanoma data (all observations).

Normally, you would be more interested in showing the Kaplan–Meier estimate graphically than numerically. To do this (Figure 14.1), you simply write

```
> plot(surv.all)
```

The markings on the curve indicate censoring times, and the bands give approximate confidence intervals. If you look closely, you will see that the bands are not symmetrical around the estimate. They are constructed as a symmetric interval on the log scale and transformed back to the original scale.

It is often useful to plot two or more survival functions on the same plot so that they can be directly compared (Figure 14.2). To obtain survival functions split by gender, do the following:

```
> surv.bysex <- survfit(Surv(days,status==1)~sex)
> plot(surv.bysex)
```

That is, you use a model formula as in lm and glm, specifying that the survival object generated from day and status should be described by sex. Notice that there are no confidence intervals on the curves. These are

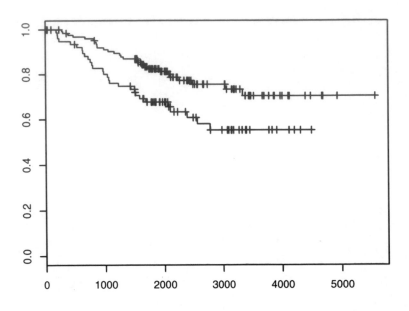

Figure 14.2. Kaplan–Meier plots for melanoma data, grouped by gender.

turned off when there are two or more curves because the display easily becomes confusing. They can be turned on again by passing `conf.int=T` to `plot`, in which case it can be recommended to use separate colours for the curves, as in

```
> plot(surv.bysex, conf.int=T, col=c("black","gray"))
```

Similarly, you can avoid plotting the confidence bands in the single-sample case by setting `conf.int=F`. If you want the bands but at a 99% confidence level, you should pass `conf.int=0.99` to `survfit`. Notice that the level of confidence is an argument to the fitting function (which needs it to compute the confidence limits), whereas the decision to plot the bands is controlled by a similarly named argument to plot.

14.4   The log-rank test

The log-rank test is used to test whether two or more survival curves are identical. It is based on looking at the population at each death time and computing the expected number of deaths in proportion to the number of individuals at risk in each group. This is then summed over all death times and compared with the observed number of deaths by a procedure similar (but not identical) to the χ^2 test. Notice that the interpretation of "expected" and "observed" is slightly peculiar: If the difference in mortality is sufficiently large, then you can easily "expect" the same individuals to die several times over the course of the trial. If the population is observed to extinction with no censoring, then the observed number of deaths will equal the group size by definition and the expected values will contain all the random variation.

The log-rank test is formally nonparametric since the distribution of the test statistic depends only on the assumption that the groups have the same survival function. However, it can also be viewed as a model-based test under the assumption of *proportional hazards* (see Section 14.1). You can set up a semiparametric model in which the hazard itself is unspecified but it is assumed that the hazards are proportional between groups. Testing that the proportionality factors are all unity then leads to a log-rank test. The log-rank test will work best against this class of alternatives.

Computation of the log-rank test is done by the function `survdiff`. This actually implements a whole family of tests specified by a parameter ρ, allowing various nonproportional hazards alternatives to the null hypothesis, but the default value of $\rho = 0$ gives the log-rank test.

```
> survdiff(Surv(days,status==1)~sex)
Call:
survdiff(formula = Surv(days, status == 1) ~ sex)

        N Observed Expected (O-E)^2/E (O-E)^2/V
sex=1 126       28     37.1      2.25      6.47
sex=2  79       29     19.9      4.21      6.47

 Chisq= 6.5  on 1 degrees of freedom, p= 0.011
```

The specification is using a model formula as for linear and generalized linear models. However, the test can deal only with grouped data, so if you specify multiple variables on the right-hand side it will work on the grouping of data generated by all combinations of predictor variables. It also makes no distinction between factors and numerical codes. The same is true of `survfit`.

It is also possible to specify stratified analyses, in which the observed and expected value calculations are carried out separately within a stratification of the data set. For instance, you can compute the log-rank test for a gender effect stratified by ulceration as follows:

```
> survdiff(Surv(days,status==1)~sex+strata(ulc))
Call:
```

```
survdiff(formula = Surv(days, status == 1) ~ sex + strata(ulc))

          N Observed Expected (O-E)^2/E (O-E)^2/V
sex=1 126       28     34.7      1.28      3.31
sex=2  79       29     22.3      1.99      3.31

 Chisq= 3.3  on 1 degrees of freedom, p= 0.0687
```

Notice that this makes the effect of sex appear less significant. A possible explanation might be that males seek treatment when the disease is in a more advanced state than women do, so that the gender difference is reduced when adjusted for a measure of disease progression.

14.5   The Cox proportional hazards model

The proportional hazards model allows the analysis of survival data by regression models similar to those of lm and glm. The scale on which linearity is assumed is the log-hazard scale. Models can be fitted via the maximization of *Cox's likelihood*, which is not a true likelihood but it can be shown that it may be used as one. It is calculated in a manner similar to that of the log-rank test, as the product of conditional likelihoods of the observed death at each death time.

As a first example, consider a model with the single regressor sex:

```
> summary(coxph(Surv(days,status==1)~sex))
Call:
coxph(formula = Surv(days, status == 1) ~ sex)

  n= 205
      coef exp(coef) se(coef)   z     p
sex 0.662      1.94    0.265 2.5 0.013

      exp(coef) exp(-coef) lower .95 upper .95
sex        1.94      0.516      1.15      3.26

Rsquare= 0.03    (max possible= 0.937 )
Likelihood ratio test= 6.15   on 1 df,    p=0.0131
Wald test               = 6.24   on 1 df,    p=0.0125
Score (logrank) test = 6.47   on 1 df,    p=0.0110
```

The coef is the estimated logarithm of the hazard ratio between the two groups, which for convenience is also given as the actual hazard ratio exp(coef). The line following that also gives the inverted ratio (swapping the groups) and confidence intervals for the hazard ratio. Finally, three overall tests for significant effects in the model are given. These are all equivalent in large samples but may differ somewhat in small-sample

cases. Notice that the Wald test is identical to the z test based on the estimated coefficient divided by its standard error, whereas the score test is
equivalent to the log-rank test (as long as the model involves only a simple
grouping).

A more elaborate example, involving a continuous covariate and a
stratification variable, is

```
> summary(coxph(Surv(days,status==1)~sex+log(thick)+strata(ulc)))
Call:
coxph(formula = Surv(days, status == 1) ~ sex + log(thick) +
    strata(ulc))

  n= 205

              coef exp(coef)  se(coef)     z      p
sex           0.36      1.43    0.270  1.33 0.1800
log(thick)    0.56      1.75    0.178  3.14 0.0017

           exp(coef) exp(-coef) lower .95 upper .95
sex             1.43      0.698     0.844      2.43
log(thick)      1.75      0.571     1.234      2.48

Rsquare= 0.063    (max possible= 0.9 )
Likelihood ratio test= 13.3  on 2 df,    p=0.00130
Wald test            = 12.9  on 2 df,    p=0.00160
Score (logrank) test = 13.0  on 2 df,    p=0.00152
```

It is seen that the significance of the sex variable has been further reduced.

The Cox model assumes an underlying baseline hazard function with a
corresponding survival curve. In a stratified analysis, there will be one
such curve for each stratum. They can be extracted by using survfit on
the output of coxph and of course be plotted using the plot method for
survfit objects (Figure 14.3):

```
> plot(survfit(coxph(Surv(days,status==1)~
+                log(thick)+sex+strata(ulc)))))
```

Be aware that the default for survfit is to generate curves for a pseudo-
individual for which the covariates are at their mean values. In the present
case, that would correspond to a tumor thickness of 1.86 mm and a gender of 1.39 (!). Notice that we have been sloppy in not defining sex as a
factor variable, but that would not actually give a different result (coxph
subtracts the means of the regressors before fitting, so a 1/2 coding is the
same as 0/1, which is what a factor with treatment contrasts gives you).
However, you can use the newdata argument of survfit to specify a
data frame for which you want to calculate survival curves.

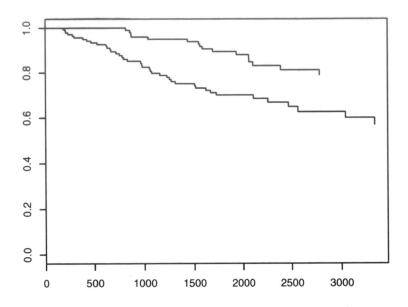

Figure 14.3. Baseline survival curves (ulcerated and nonulcerated tumors) in stratified Cox regression.

14.6   Exercises

14.1   In the `graft.vs.host` data set, estimate the survival function for patients with or without GVHD. Test the hypothesis that the survival is the same in both groups. Extend the analysis by including the other explanatory variables.

14.2   With the Cox model in the last section of the text, generate a plot with estimated survival curves for men with nonulcerated tumors of thicknesses 0.1, 0.2, and 0.5 mm (three curves in one plot). Hint: `survfit` objects can be indexed with `[ ]` to extract individual strata.

14.3   Fit Cox models to the `stroke` data with `age` and `sex` as predictors and with `sex` alone. Explain the difference.

14.4   With the split data from Exercise 10.4, you can fit a Cox model with *delayed entry* to the stroke data; `help(Surv)` shows how to set up the `Surv` object in that case. Refit the model(s) from the previous exercise.

15

Rates and Poisson regression

Epidemiological studies often involve the calculation of rates, typically rates of death or incidence rates of a chronic or acute disease. This is based upon counts of events occurring within a certain amount of time. The Poisson regression method is often employed for the statistical analysis of such data. However, data that are not actually counts of events but rather measurements of time until an event (or nonevent) can be analyzed by a technique which is formally equivalent.

15.1   Basic ideas

The data that we wish to analyze can be in one of two forms. They can be in *aggregate* form as an observed count x based on a number of person-years T. Often the latter is an approximation based on tables of population size. There may of course be more than one group, and we may wish to formulate various models describing the rates in different groups.

We may also have individual-level data, in which for each subject we have a time under observation T_i and a 0/1 indicator x_i of whether the subject has had an event. The aggregate data can be thought of as being $x = \sum x_i$ and $T = \sum T_i$, where the sums are over all individuals in the group.

P. Dalgaard, *Introductory Statistics with R*,
DOI: 10.1007/978-0-387-79054-1_15, © Springer Science+Business Media, LLC 2008

15.1.1 The Poisson distribution

The Poisson distribution can be described as the limiting case of the binomial distributions when the size parameter N increases while the expected number of successes $\lambda = Np$ is fixed. This is useful to describe rare event in large populations. The resulting distribution has point probabilities

$$f(x) = \frac{\lambda^x}{x!} e^{-\lambda} \qquad x = 0, 1, \ldots$$

The distribution is theoretically unbounded, although the probabilities for large x will be very small. In R, the Poisson distribution is available via the functions dpois, ppois, etc.

In the context of epidemiological data, the parameter of interest is usually the expected counts *per unit of observed time*; i.e., the rate at which events occur. This enables comparison of populations that may be of different size or observed for different lengths of time. Accordingly, we may parameterize the Poisson distribution using

$$\rho = \lambda / T$$

Notice that parts of the literature use λ to denote the rate. The notation used here is chosen so as to stay compatible with the argument name in dpois.

The Poisson likelihood

Models for Poisson data can be fitted by the method of maximum likelihood. If we parameterize in terms of ρ, the log-likelihood becomes

$$l(\rho) = \text{constant} + x \log \rho - \rho T$$

which is maximized when $\rho = x/T$. The log-likelihood can be generalized to models involving several counts by summing terms of the same form.

15.1.2 Survival analysis with constant hazard

In this section, for convenience, we use terminology appropriate for mortality studies, although the event may be many things other than the death of the subject.

Individual-level data are essentially survival data as described in Chapter 14, except for changes in notation. One difference, though, is that in the analysis of rates it is often reasonable to assume that the hazard does not change over time, or at least not abruptly so. Rates tend to be obtained over rather short individual time periods, and the origin of the timescale is not usually keyed to a life-changing event such as disease onset or major surgery.

If the hazard is constant, then the distribution of the lifetime is the *exponential distribution* with density $\rho e^{-\rho t}$ and survival function $e^{-\rho t}$.

Likelihood analysis

Likelihoods for censored data can be constructed using terms that are either the probability density at the time of death or the survival probability in the case of censoring. In the constant-hazard case, the two kinds of terms differ only in the presence of the factor ρ, which we may conveniently encode using the event indicator x_i so that the log-likelihood terms are

$$l(\rho) = x_i \log \rho - \rho T_i$$

Except for the constant, which does not depend on ρ, these terms are formally identical to a Poisson likelihood, where the count is 1 (death) or zero (censoring). This is the crucial "trick" that allows survival data with constant hazard to be analyzed by Poisson regression methods.

The trick can be extended to hazards that are only piecewise constant. Suppose the lifetime of an individual is subdivided as $T_i = T_i^{(1)} + \cdots + T_i^{(k)}$, where the hazard is assumed constant during each section of time. The corresponding log-likelihood term is

$$l(\rho_1, \ldots, \rho_k) = \sum_{j=1}^{k} (x_i^{(j)} \log \rho_j - \rho_j T_i^{(j)})$$

in which the first $k - 1$ of the $x_i^{(j)}$ will be 0, and only the last one, $x_i^{(k)}$, can be either 0 or one. The point of writing it in this elaborate form is that it then becomes obvious that the likelihood contribution *might as well* have come from k different individuals where the first $k - 1$ had censored observations.

This is the rationale behind time-splitting techniques where the observation time of one subject is divided into observations for multiple pseudo-individuals.

It should be noted that although the models with (piecewise) constant hazard can be fitted and analyzed by likelihood techniques, pretending that the data have come from a Poisson distribution, this does not extend to all aspects of the model. For instance following a cohort to extinction will lead to a fixed total number of events by definition, whereas the corresponding Poisson model implies that the total event count has a Poisson distribution. Both types of models deal in rates, counts per time, but the difference is to what extent the random variation lies in the counts or in the amount of time. When data are frequently censored (i.e., the event is rare), the survival model becomes well approximated by the Poisson model.

15.2   Fitting Poisson models

The class of generalized linear models (see Section 13.1) also includes the Poisson distribution, which by default uses a log link function. This is the mathematically convenient option and also a quite natural choice since it allows the linear predictor to span the entire real line. We can use this to formulate models for the log rates of the form

$$\log \rho = \beta_0 + \beta_1 x_1 + \beta_2 x_2 + \dots \beta_k x_k$$

or, since glm needs a model for the expected counts rather than rates,

$$\log \lambda = \beta_0 + \beta_1 x_1 + \beta_2 x_2 + \dots \beta_k x_k + \log T$$

A feature of many Poisson models is that the model contains an *offset* in the linear predictor, $\log T$ in this case. Notice that this is not the same as including the term as a regression variable since the regression coefficient is fixed at 1.

The following example was used by Erling B. Andersen in 1977. It involves the rates of lung cancer by age in four Danish cities and may be found as eba1977 in the ISwR package.

```
> names(eba1977)
[1] "city"   "age"    "pop"    "cases"
> attach(eba1977)
```

To fit a model that has multiplicative effects of age and city on the rate of lung cancer cases, we use the glm function in much the same way as in logistic regression. Of course, we need to change the family argument to accommodate Poisson-distributed data. We also need to incorporate an offset to account for the different sizes and age structures of the populations in the four cities.

```
> fit <- glm(cases~city+age+offset(log(pop)), family=poisson)
> summary(fit)
Call:
glm(formula = cases ~ city + age + offset(log(pop)), family=poisson)

Deviance Residuals:
     Min        1Q     Median         3Q        Max
-2.63573   -0.67296   -0.03436    0.37258    1.85267
```

```
Coefficients:
             Estimate Std. Error z value Pr(>|z|)
(Intercept)   -5.6321     0.2003 -28.125  < 2e-16 ***
cityHorsens   -0.3301     0.1815  -1.818   0.0690 .
cityKolding   -0.3715     0.1878  -1.978   0.0479 *
cityVejle     -0.2723     0.1879  -1.450   0.1472
age55-59       1.1010     0.2483   4.434 9.23e-06 ***
age60-64       1.5186     0.2316   6.556 5.53e-11 ***
age65-69       1.7677     0.2294   7.704 1.31e-14 ***
age70-74       1.8569     0.2353   7.891 3.00e-15 ***
age75+         1.4197     0.2503   5.672 1.41e-08 ***
---
Signif. codes:  0 '***' 0.001 '**' 0.01 '*' 0.05 '.' 0.1 ' ' 1

(Dispersion parameter for poisson family taken to be 1)

    Null deviance: 129.908  on 23  degrees of freedom
Residual deviance:  23.447  on 15  degrees of freedom
AIC: 137.84

Number of Fisher Scoring iterations: 5
```

The offset was included in the model formula in this case. Alternatively, it could have been given as a separate argument as in

```
glm(cases~city+age, offset = log(pop), family=poisson)
```

The table labelled "Coefficients:" contains regression coefficients for the linear predictor along with standard errors and z tests. These can be interpreted in the same way as in ordinary multiple regression or logistic regression. Since both variables are factors and we are using treatment contrasts (see Section 12.3), the coefficients indicate differences in the log rate (i.e., the log of the rate ratio) compared with the city of Fredericia and with the 50–54-year-olds, respectively.

The intercept term refers to the log rate for the group of 50–54-year-olds in Fredericia. Notice that because we used the population size rather than the number of person-years in the offset and the data cover the years 1968–1971, this rate will effectively be per 4 person-years.

A goodness-of-fit statistic is provided by comparing the residual deviance to a χ^2 distribution on the stated degrees of freedom. This statistic is generally considered valid if the expected count in all cells is larger than 5. Accordingly,

```
> min(fitted(fit))
[1] 6.731286
> pchisq(deviance(fit), df.residual(fit), lower=F)
[1] 0.07509017
```

and we see that the model fits the data acceptably. Of course, we could also just have read off the residual deviance and degrees of freedom from the summary output:

```
> pchisq(23.45, 15, lower=F)
[1] 0.07504166
```

From the coefficient table, it is obvious that there is an age effect, but it is less clear whether there is a city effect. We can perform χ^2 tests for each term by using drop1 and looking at the changes in the deviance.

```
> drop1(fit, test="Chisq")
Single term deletions

Model:
cases ~ city + age + offset(log(pop))
        Df Deviance     AIC      LRT Pr(Chi)
<none>        23.447 137.836
city     3    28.307 136.695    4.859  0.1824
age      5   126.515 230.903  103.068   <2e-16 ***
...
```

We see that the age term is significant, hardly surprisingly, but the city term apparently is not. However, if you can argue a priori that Fredericia could be expected to have a higher cancer rate than the three other cities, then it could be warranted to combine the three other cities into one and perform an analysis as below.

```
> fit2 <- glm(cases~(city=="Fredericia")+age+offset(log(pop)),
+                    family=poisson)
> anova(fit, fit2, test="Chisq")
Analysis of Deviance Table

Model 1: cases ~ city + age + offset(log(pop))
Model 2: cases ~ (city == "Fredericia") + age + offset(log(pop))
  Resid. Df Resid. Dev Df Deviance P(>|Chi|)
1        15    23.4475
2        17    23.7001 -2  -0.2526    0.8814
> drop1(fit2, test="Chisq")
Single term deletions

Model:
cases ~ (city == "Fredericia") + age + offset(log(pop))
                      Df Deviance     AIC      LRT Pr(Chi)
<none>                      23.700 134.088
city == "Fredericia"  1    28.307 136.695    4.606 0.03185 *
age                   5   127.117 227.505  103.417 < 2e-16 ***
...
```

According to this, you may combine the three cities other than Fredericia, and, once this is done, Fredericia does indeed appear to be significantly

different from the others. Alternatively, you can look at the coefficients in
`fit2` directly

```
> summary(fit2)
...
Coefficients:
                        Estimate Std. Error z value Pr(>|z|)
(Intercept)              -5.9589     0.1809 -32.947  < 2e-16 ***
city == "Fredericia"TRUE  0.3257     0.1481   2.200   0.0278 *
age55-59                  1.1013     0.2483   4.436 9.17e-06 ***
age60-64                  1.5203     0.2316   6.564 5.23e-11 ***
age65-69                  1.7687     0.2294   7.712 1.24e-14 ***
age70-74                  1.8592     0.2352   7.904 2.71e-15 ***
age75+                    1.4212     0.2502   5.680 1.34e-08 ***
...
```

and see the p-value of 0.0278. This agrees with the 0.03185 from `drop1`;
you cannot expect the two p-values to be perfectly equal since they rely
on different asymptotic approximations. If you really push it, you can argue that a one-sided test with half the p-value is appropriate since you
would only expect Fredericia to be more harmful than the others, not less.
However, the argumentation becomes tenuous, and in his paper Andersen outlines the possibility of testing Fredericia against the other cities but
stops short of providing any p-value, stating that in his opinion "there is
no reason to believe a priori that Fredericia is the more dangerous city".

It is sometimes preferred to state the results of Poisson regression analysis in terms of *rate ratios* by taking `exp()` of the estimates (this parallels
the presentation of logistic regression analysis in terms of odds ratios in
Section 13.4). The intercept term is not really a ratio but a rate, and for
nonfactor covariates it should be understood that the coefficient is the relative change *per unit* change in the covariate. Because of the nonlinear
transformation, standard errors are not useful; instead one can calculate
confidence intervals for the coefficients as follows:

```
> cf <- coefficients(summary(fit2))
> est <- cf[,1]
> s.e. <- cf[,2]
> rr <- exp(cbind(est, est - s.e.*qnorm(.975), est
+                     + s.e.*qnorm(.975) ))
> colnames(rr) <- c("RateRatio", "CI.lo","CI.hi")
> rr
                          RateRatio        CI.lo        CI.hi
(Intercept)              0.002582626 0.001811788  0.003681423
city == "Fredericia"TRUE 1.384992752 1.036131057  1.851314957
age55-59                 3.008134852 1.849135187  4.893571521
age60-64                 4.573665854 2.904833526  7.201245496
age65-69                 5.863391064 3.740395488  9.191368903
age70-74                 6.418715646 4.047748963 10.178474731
age75+                   4.142034525 2.536571645  6.763637070
```

Actually, we can do better by using the `confint` function. This calculates confidence intervals by profiling the likelihood function instead of using the approximation with the normal distribution inherent in the use of asymptotic standard errors. This is done like this:

```
> exp(cbind(coef(fit2), confint(fit2)))
Waiting for profiling to be done...
                                            2.5 %       97.5 %
(Intercept)               0.002582626 0.001776461  0.003617228
city == "Fredericia"TRUE  1.384992752 1.029362341  1.841224091
age55-59                  3.008134852 1.843578634  4.902339637
age60-64                  4.573665854 2.912314045  7.248143959
age65-69                  5.863391064 3.752718226  9.256907108
age70-74                  6.418715646 4.053262281 10.234338998
age75+                    4.142034525 2.527117848  6.771833979
```

In the present case, we are well within the regime where the asymptotic normal approximation works well, so there is little difference between the two displays. However, in some cases where some expected cell counts are low and one or several coefficients are poorly determined, the difference can be substantial.

15.3   Computing rates

We return to the Welsh nickel worker data discussed in Chapter 10. In that section, we discussed how to split the individual lifetime data into smaller pieces that could reasonably be merged with the standard mortality table in the `ewrates` data.

The result of this initial data restructuring is in the `nickel.expand` data set. It contains data from a lot of short time intervals like this:

```
> head(nickel.expand)
  agr  ygr  id icd exposure      dob   age1st    agein ageout lung
1  20 1931 325   0        0 1910.500 14.0737 23.7465     25    6
2  20 1931 273   0        0 1909.500 14.6913 24.7465     25    6
3  20 1931 110   0        0 1909.247 14.0302 24.9999     25    6
4  20 1931 574   0        0 1909.729 14.0356 24.5177     25    6
5  20 1931 213   0        0 1910.129 14.2018 24.1177     25    6
6  20 1931 546   0        0 1909.500 14.4945 24.7465     25    6
  nasal other
1     0  3116
2     0  3116
3     0  3116
4     0  3116
5     0  3116
6     0  3116
```

The same individuals reappear later in the data at older ages. For example, all data for the individual with id number 325 are

```
> subset(nickel.expand, id==325)
     agr  ygr  id icd exposure    dob  age1st    agein   ageout lung
1     20 1931 325   0        0 1910.5 14.0737 23.7465 25.0000    6
13    25 1931 325   0        0 1910.5 14.0737 25.0000 30.0000   14
172   30 1936 325   0        0 1910.5 14.0737 30.0000 35.0000   30
391   35 1941 325   0        0 1910.5 14.0737 35.0000 40.0000   81
728   40 1946 325 434        0 1910.5 14.0737 40.0000 43.0343  236
    nasal other
1       0  3116
13      0  3024
172     1  3188
391     1  3549
728     3  3643
```

Accordingly, this subject enters the study at age 23.7 and we follow him through five age groups until his death at age 43.

The variable ygr reflects the year of entry into the interval, so even though the subject dies in 1953, the last record is coded as belonging to the years 1946–1950.

Subject no. 325 has the icd code 434 in his last record. This refers to the International Classification of Diseases (version 7) and indicates "Other and unspecified diseases of the heart" as the cause of death. For the purposes of this chapter, we are primarily interested in lung cancer, which has codes 162 and 163, so we define a variable to indicate whether this is the cause of death. (Expect a warning about masking the lung data set upon attaching.)

```
> nickel.expand <- within(nickel.expand,
+     lung.cancer <- as.numeric(icd %in% c(162,163)))
> attach(nickel.expand)
```

The %in% operator returns a logical vector that is TRUE when the corresponding element of the operand on the left is contained in the vector that is the operand on the right and FALSE in all other cases. Use of this operator is slightly dangerous in the case of an NA element in icd, but in these particular data, there are none. We convert the result to zero or one since we are going to pretend that it is a Poisson count later on (this is not strictly necessary). Notice that by using lung.cancer as the endpoint, we treat death from all other causes, including "unknown", as censoring.

Each record provides ageout − agein person-years of risk time, so to tabulate the risk times, we can just do as follows:

```
> pyr <- tapply(ageout-agein,list(ygr,agr), sum)
> print(round(pyr), na.print="-")
```

```
       20 25  30  35  40  45  50  55  60  65  70  75 80
1931    3 86 268 446 446 431 455 323 159  23   4   –  –
1936    –  – 100 327 504 512 503 472 314 130  20   5  –
1941    –  –   0 105 336 481 482 445 368 235  80  14  3
1946    –  –   –   – 102 335 461 404 369 263 157  43 10
1951    –  –   –   –   –  95 299 415 334 277 181  92 31
1956    –  –   –   –   –   –  89 252 364 257 181 101 52
1961    –  –   –   –   –   –   –  71 221 284 150 104 44
1966    –  –   –   –   –   –   –   –  66 168 208  93 51
1971    –  –   –   –   –   –   –   –   –  57 133 131 54
1976    –  –   –   –   –   –   –   –   –   –  31  68 53
```

Notice that there are many NA entries in cells that no subject ever entered. The subjects in the study were born between 1864 and 1910, so there is a large block missing in the lower left and a smaller block in the upper right. The na.print option to print allows you to represent these missing values by a string that is less visually imposing than the default "NA".

The corresponding counts of lung cancer cases are obtained as

```
> count <- tapply(lung.cancer, list(ygr, agr), sum)
> print(count, na.print="-")
     20 25 30 35 40 45 50 55 60 65 70 75 80
1931  0  0  0  0  0  4  2  2  2  0  0  -  -
1936  -  -  0  0  2  3  4  6  5  1  0  0  -
1941  -  -  0  0  0  3  7  5  6  3  2  0  0
1946  -  -  -  -  0  0  8  7  6  2  2  0  0
1951  -  -  -  -  -  0  3  3  9  6  1  0  0
1956  -  -  -  -  -  -  0  4  3  6  1  2  0
1961  -  -  -  -  -  -  -  0  1  1  3  2  1
1966  -  -  -  -  -  -  -  -  2  0  0  1  0
1971  -  -  -  -  -  -  -  -  -  0  0  2  2
1976  -  -  -  -  -  -  -  -  -  -  0  1  1
```

and the cancer rates can be obtained as the ratio of the counts to the risk time. These are small, so we multiply by 1000 to get rates per 1000 person-years.

```
> print(round(count/pyr*1000, 1), na.print="-")
     20 25 30 35 40  45   50   55   60   65   70   75   80
1931  0  0  0  0  0 9.3  4.4  6.2 12.6  0.0  0.0    -    -
1936  -  -  0  0  4 5.9  7.9 12.7 15.9  7.7  0.0  0.0    -
1941  -  -  0  0  0 6.2 14.5 11.2 16.3 12.8 25.0  0.0  0.0
1946  -  -  -  -  0 0.0 17.4 17.3 16.3  7.6 12.8  0.0  0.0
1951  -  -  -  -  - 0.0 10.0  7.2 27.0 21.7  5.5  0.0  0.0
1956  -  -  -  -  -   -  0.0 15.9  8.2 23.4  5.5 19.8  0.0
1961  -  -  -  -  -   -    -  0.0  4.5  3.5 19.9 19.3 22.8
1966  -  -  -  -  -   -    -    - 30.1  0.0  0.0 10.7  0.0
1971  -  -  -  -  -   -    -    -    -  0.0  0.0 15.2 36.8
1976  -  -  -  -  -   -    -    -    -    -  0.0 14.6 19.0
```

Comparison of these rates with those in `ewrates` suggests that they are very high. However, this kind of display has the disadvantage that it hides the actual counts on which the rates are based. For instance, the lower part of the column for 80–84-year-olds jumps by roughly 20 units for each additional case since there are only about 50 person-years per cell.

It may be better to compute the expected counts in each cell based on the standard mortality table and then compare that to the actual counts. Since we have already merged in the `ewrates` data, this is just a matter of multiplying each piece of risk time by the rate. We need to divide by 1e6 (i.e., $10^6 = 1000000$) since the standard rates are given per million person-years.

```
> expect.count <- tapply(lung/1e6*(ageout-agein),
+                        list(ygr,agr), sum)
> print(round(expect.count, 1), na.print="-")
     20 25 30 35  40  45  50  55  60  65  70  75  80
1931  0  0  0   0 0.1 0.1 0.2 0.2 0.1 0.0 0.0   -   -
1936  -  -  0   0 0.1 0.1 0.2 0.3 0.2 0.1 0.0 0.0   -
1941  -  -  0   0 0.1 0.2 0.3 0.4 0.4 0.2 0.1 0.0 0.0
1946  -  -  -   - 0.0 0.2 0.4 0.5 0.6 0.5 0.2 0.0 0.0
1951  -  -  -   -   - 0.1 0.4 0.8 0.9 0.8 0.5 0.2 0.0
1956  -  -  -   -   -   - 0.1 0.6 1.2 1.0 0.7 0.3 0.1
1961  -  -  -   -   -   -   - 0.2 0.8 1.4 0.7 0.5 0.1
1966  -  -  -   -   -   -   -   - 0.2 0.9 1.3 0.6 0.2
1971  -  -  -   -   -   -   -   -   - 0.3 0.9 1.0 0.3
1976  -  -  -   -   -   -   -   -   -   - 0.2 0.6 0.4
```

The observed counts are clearly much larger than expected. We can summarize them by calculating the overall SMR (standardized mortality rate), which is simply the ratio of the total number of cases to the total expected number of cases.

```
> expect.tot <- sum(lung/1e6*(ageout-agein))
> expect.tot
[1] 24.19893
> count.tot <- sum(lung.cancer)
> count.tot
[1] 137
> count.tot/expect.tot
[1] 5.661408
```

That is, this data set has almost six times as many cancer deaths as you would expect from the mortality of the general population.

15.4   Models with piecewise constant intensities

We can formulate the SMR analysis as a "Poisson" regression model in the sense of Section 15.1.2. The assumption behind the SMR is that there is a constant rate ratio to the standard mortality, so we can fit a model with only an intercept while having an offset, which is the log of the expected count. This is not really different from modelling rates — the population mortality ρ_i is just absorbed into the offset, $\log \rho_i + \log T_i = \log \rho_i T_i$.

```
> fit <- glm(lung.cancer ~ 1, poisson,
+              offset = log((ageout-agein)*lung/1e6))
> summary(fit)
...
Coefficients:
            Estimate Std. Error z value Pr(>|z|)
(Intercept)  1.73367    0.08544   20.29   <2e-16 ***
---
Signif. codes:  0 '***' 0.001 '**' 0.01 '*' 0.05 '.' 0.1 ' ' 1

(Dispersion parameter for poisson family taken to be 1)

    Null deviance: 1175.6  on 3723   degrees of freedom
Residual deviance: 1175.6  on 3723   degrees of freedom
AIC: 1451.6

Number of Fisher Scoring iterations: 7
```

Notice that this is based on individual data; the dependent variable lung.cancer is zero or one. We could have aggregated the data according to the cross-classification of agr and ygr and analyzed the number of cases in each cell. This would have allowed glm to run much faster, but on the other hand it would then not be possible to add individual covariates such as age at first exposure.

In this case, we cannot use the deviances for model checking both because the expected counts per cell are very small and because we do not actually have Poisson-distributed data. However, the standard error and the p-value should be reliable if the assumptions hold.

The connection between this analysis and the SMR can be seen immediately from

```
> exp(coef(fit))
(Intercept)
   5.661408
```

This value is exactly the SMR value from the previous section.

We can analyze the data more thoroughly using regression methods. As a first approach, we investigate whether the SMR is constant over year and age groups using a multiplicative Poisson model.

We need to simplify the groupings because some of the groups contain very few cases. By calculating the marginal tables of counts, we get some idea of what to do.

```
> tapply(lung.cancer, agr, sum)
20 25 30 35 40 45 50 55 60 65 70 75 80
 0  0  0  0  2 10 24 27 34 19  9  8  4
> tapply(lung.cancer, ygr, sum)
1931 1936 1941 1946 1951 1956 1961 1966 1971 1976
  10   21   26   25   22   16    8    3    4    2
```

To get at least 10 cases per level, we combine all values of agr up to 45 (i.e., ages less than 50) and also those from 70 and up. Similarly, we combine all values of ygr for the periods from 1961 onwards.

```
> detach()
> nickel.expand <- within(nickel.expand, {
+      A <- factor(agr)
+      Y <- factor(ygr)
+      lv <- levels(A)
+      lv[1:6] <- "< 50"
+      lv[11:13] <- "70+"
+      levels(A) <- lv
+      lv <- levels(Y)
+      lv[7:10] <- "1961ff"
+      levels(Y) <- lv
+      rm(lv)
+ })
> attach(nickel.expand)
```

Notice that this is a case where the within function (see Section 2.1.8) works better than transform because it allows more flexibility, including the creation of temporary variables such as lv.

We can analyze the effect of A and Y on the mortality ratio by building a log-additive model in the usual way. Notice that we still use the original grouping in the calculation of the offset; it is only the SMR that is assumed to be the same for everyone below 50, etc. We use drop1 to test the significance of the two factors.

```
> fit <- glm(lung.cancer ~ A + Y, poisson,
+             offset=log((ageout-agein)*lung/1e6))
> drop1(fit, test="Chisq")
Single term deletions

Model:
lung.cancer ~ A + Y
```

```
       Df Deviance      AIC     LRT  Pr(Chi)
<none>      1069.73 1367.73
A       5  1073.81 1361.81    4.08   0.5376
Y       6  1118.50 1404.50   48.77 8.29e-09 ***
---
Signif. codes:  0 '***' 0.001 '**' 0.01 '*' 0.05 '.' 0.1 ' ' 1
```

So it seems that we do not need the age grouping in the model, but
the year grouping is needed. Accordingly, we fit a model with Y alone,
and by dropping the intercept, we get a parameterization with a separate
intercept for each level of Y.

```
> fit <- glm(lung.cancer ~ Y - 1, poisson,
+               offset=log((ageout-agein)*lung/1e6))
> summary(fit)
...
Coefficients:
          Estimate Std. Error z value Pr(>|z|)
Y1931       2.6178     0.3162   8.279  < 2e-16 ***
Y1936       3.0126     0.2182  13.805  < 2e-16 ***
Y1941       2.7814     0.1961  14.182  < 2e-16 ***
Y1946       2.2787     0.2000  11.394  < 2e-16 ***
Y1951       1.8038     0.2132   8.461  < 2e-16 ***
Y1956       1.3698     0.2500   5.479 4.27e-08 ***
Y1961ff     0.4746     0.2425   1.957   0.0504 .
....
```

The regression coefficients may again be recognized as log-SMR values, as
the following demonstrates:

```
> round(exp(coef(fit)), 1)
   Y1931    Y1936    Y1941    Y1946    Y1951    Y1956 Y1961ff
    13.7     20.3     16.1      9.8      6.1      3.9     1.6
> expect.count <-  tapply(lung/1e6*(ageout-agein), Y, sum)
> count <- tapply(lung.cancer, Y, sum)
> cbind(count=count, expect=round(expect.count,1),
+       SMR= round(count/expect.count, 1))
        count expect  SMR
1931       10    0.7 13.7
1936       21    1.0 20.3
1941       26    1.6 16.1
1946       25    2.6  9.8
1951       22    3.6  6.1
1956       16    4.1  3.9
1961ff     17   10.6  1.6
```

The advantage of using the regression approach is that it provides a framework in which you can formulate statistical tests and investigate the effect
of multiple regression variables simultaneously.

Breslow and Day analyzed the nickel data in their seminal book (Breslow and Day, 1987) on the analysis of cohort studies. In their analysis, they split the individual risk times according to three criteria, two of them being age and period, to match the standard mortality table, but they also treat time from employment as a time-dependent covariate with a piecewise constant effect, which requires that the person-year be split further according to the interval boundaries. They then represent time effects using three variables: time since, age at, and year of first employment, TFE, AFE, and YFE, respectively. In addition, they include a measure of exposure level.

The following analysis roughly reproduces the Breslow and Day analysis. It is not completely similar because we settle for splitting time according to agr only and use the age at entry into each interval to define the TFE variable as well as for choosing the relevant standard mortality rates. However, to enable some comparison of results, we define cut groups in a manner that is similar to that of Breslow and Day.

```
> detach()
> nickel.expand <- within(nickel.expand,{
+      TFE <- cut(agein-agelst, c(0,20,30,40,50,100), right=F)
+      AFE <- cut(agelst, c(0, 20, 27.5, 35, 100), right=F)
+      YFE <- cut(dob + agelst, c(0, 1910, 1915, 1920, 1925),right=F)
+      EXP <- cut(exposure, c(0, 0.5, 4.5, 8.5, 12.5, 25), right=F)
+ })
> attach(nickel.expand)
```

Some relabelling of group levels might be called for — e.g., the levels for EXP are really 0, 0.5–4, 4.5–8, 8.5–12, 12.5+ — but let us not make more of it than necessary.

We fit a multiplicative model and test the significance of the individual terms as follows:

```
> fit <- glm(lung.cancer ~ TFE + AFE + YFE + EXP, poisson,
+            offset=log((ageout-agein)*lung/1e6))
> drop1(fit, test="Chisq")
Single term deletions
```

```
Model:
lung.cancer ~ TFE + AFE + YFE + EXP
        Df Deviance    AIC     LRT    Pr(Chi)
<none>        1052.91 1356.91
TFE      4   1107.33 1403.33  54.43  4.287e-11 ***
AFE      3   1054.99 1352.99   2.08  0.5560839
YFE      3   1058.06 1356.06   5.15  0.1608219
EXP      4   1071.98 1367.98  19.07  0.0007606 ***
```

This suggests that the two major terms are TFE and EXP, whereas AFE and YFE could be taken out of the model. Notice, though, that it cannot be concluded from the above that both can be removed. In principle, one of them could become significant when the other is removed. This does not happen in this case, though.

The table of coefficients looks like this:

```
> summary(fit)
...
Coefficients:
                  Estimate Std. Error z value Pr(>|z|)
(Intercept)        2.36836    0.55716   4.251 2.13e-05 ***
TFE[20,30)        -0.21788    0.36022  -0.605 0.545284
TFE[30,40)        -0.77184    0.36529  -2.113 0.034605 *
TFE[40,50)        -1.87583    0.41707  -4.498 6.87e-06 ***
TFE[50,100)       -2.22142    0.55068  -4.034 5.48e-05 ***
AFE[20,27.5)       0.28506    0.31524   0.904 0.365868
AFE[27.5,35)       0.21961    0.34011   0.646 0.518462
AFE[35,100)       -0.10818    0.44412  -0.244 0.807556
YFE[1910,1915)     0.04826    0.27193   0.177 0.859137
YFE[1915,1920)    -0.56397    0.37585  -1.501 0.133483
YFE[1920,1925)    -0.42520    0.30017  -1.417 0.156614
EXP[0.5,4.5)       0.58373    0.21200   2.753 0.005897 **
EXP[4.5,8.5)       1.03175    0.28364   3.638 0.000275 ***
EXP[8.5,12.5)      1.18345    0.37406   3.164 0.001557 **
EXP[12.5,25)       1.28601    0.48236   2.666 0.007674 **
...
```

A dose-response pattern and a declining effect of time since first employment seem to be present.

The results may be more readily interpreted if they are given in terms of ratios and confidence intervals. These can be obtained in exactly the same way as in the analysis of the eba1977 data.

15.5   Exercises

15.1   In the bcmort data set, we defined the period and area factors in Exercise 10.2. Fit a Poisson regression model to the data with age, period, and area as descriptors, as well as the three two-factor interaction terms. The interaction between period and area can be interpreted as the effect of screening.

15.2   With the split stroke data from Exercise 10.4, fit a Poisson regression model corresponding to a constant hazard in each interval and with multiplicative effects of age and sex.

16

Nonlinear curve fitting

Curve fitting problems occur in many scientific areas. The typical case is that you wish to fit the relation between some response y and a one-dimensional predictor x, by adjusting a (possibly multidimensional) parameter β. That is,

$$y = f(x; \beta) + \text{error}$$

in which the "error" term is usually assumed to contain independent normally distributed terms with a constant standard deviation σ. The class of models can be easily extended to multivariate x and somewhat less easily to models with nonconstant error variation, but we settle for the simple case.

Chapter 6 described the special case of a linear relation

$$y = \beta_0 + \beta_1 x + \text{error}$$

and we discussed the fitting of polynomials by including quadratic and higher-order terms in Section 12.1. There are other techniques, notably trigonometric regression and spline regression, that can also be formulated in linear form and handled by software for multiple regression analysis like `lm`.

However, sometimes linear methods are inadequate. The common case is that you have a priori knowledge of the form of the function. This may come from theoretical analysis of an underlying physical and chemical

P. Dalgaard, *Introductory Statistics with R*,
DOI: 10.1007/978-0-387-79054-1_16, © Springer Science+Business Media, LLC 2008

system, and the parameters of the relation have a specific meaning in that theory.

The method of least squares makes good sense even when the relation between data and parameters is not linear. That is, we can estimate β by minimizing

$$\mathrm{SSD}(\beta) = \sum(y - f(x; \beta))^2$$

There is no explicit formula for the location of the minimum, but the minimization can be performed numerically by algorithms that we describe only superficially here. This general technique is also known as nonlinear regression analysis. For an in-depth treatment of the topic, a standard reference is Bates and Watts (1988).

If the model is "well-behaved" (to use a deliberately vague term), then the model can be approximated by a linear model in the vicinity of the optimum, and it then makes sense to calculate approximate standard errors for the parameter estimates.

Most of the available optimization algorithms build on the same idea of linearization; i.e.,

$$y - f(x; \beta + \delta) \approx y - f(x; \beta) + Df\delta$$

in which Df denotes the *gradient matrix* of derivatives of f with respect to β. This effectively becomes a design matrix of a linear model, and you can proceed from a starting guess at β to find an approximate least squares fit of δ. Then you replace β by $\beta + \delta$ and repeat until convergence. Variations on this basic algorithm include numerical computation of Df and techniques to avoid instability if the starting guess is too far from the optimum.

To perform the optimization in R, you can use the nls function, which is broadly similar to lm and glm.

16.1   Basic usage

In this section, we use a simulated data set just so that we know what we are doing. The model is a simple exponential decay.

```
> t <- 0:10
> y <- rnorm(11, mean=5*exp(-t/5), sd=.2)
> plot(y ~ t)
```

The simulated data can be seen in Figure 16.1.

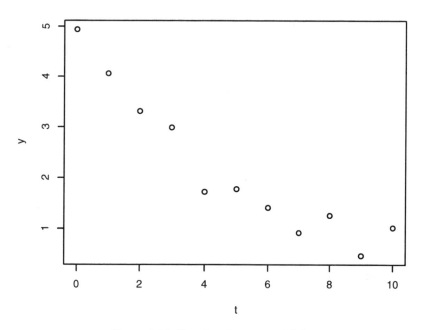

Figure 16.1. Simulated exponential decay.

We now fit the model to data using `nls`. Unlike `lm` and `glm`, the model formula for `nls` does *not* use the special codings for linear terms, grouping factors, interactions, etc. Instead, the right-hand side is an explicit expression to calculate the expected value of the left-hand side. This can depend on external variables as well as the parameters, so we need to specify which is which. The simplest way to do this is to specify a named vector (or a named list) of starting values.

```
> nlsout <- nls(y ~ A*exp(-alpha*t), start=c(A=2, alpha=0.05))
> summary(nlsout)

Formula: y ~ A * exp(-alpha * t)

Parameters:
      Estimate Std. Error t value Pr(>|t|)
A      4.97204    0.21766   22.84 2.80e-09 ***
alpha  0.20793    0.01572   13.23 3.35e-07 ***
---
Signif. codes:  0 '***' 0.001 '**' 0.01 '*' 0.05 '.' 0.1 ' ' 1

Residual standard error: 0.2805 on 9 degrees of freedom

Number of iterations to convergence: 5
```

```
Achieved convergence tolerance: 2.223e-06
```

Notice that `nls` treats t as a variable and not a parameter because it is not mentioned in the `start` argument. Whenever the fitting algorithm needs to evaluate `A * exp(-alpha * t)`, t is taken from the variable in the global environment, whereas A and `alpha` are varied by the algorithm.

The general form of the output is quite similar to that of `glm`, so we shall not dwell too long upon it. One thing that might be noted is that the t test and p-value stated for each parameter are tests for a hypothesis that the parameter is *zero*, which is often quite meaningless for nonlinear models.

16.2   Finding starting values

In the previous section, we had quite fast convergence, even though the initial guess of parameters was (deliberately) rather badly off. Unfortunately, things are not always that simple; convergence of nonlinear models can depend critically on having good starting values. Even when the algorithm is fairly robust, we at least need to get the order of magnitude right.

Methods for obtaining starting values will most often rely on an analysis of the functional form; common techniques involve transformation to linearity and the estimation of "landmarks" such asymptotes, maximum points, and initial slopes.

To illustrate this, we again consider the Juul data. This time we focus on the relation between age and height. To obtain a reasonably homogeneous data set, we look at males only and subset the data to the ages between 5 and 20.

```
> attach(subset(juul2, age<20 & age>5 & sex==1))
> plot(height ~ age)
```

A plot of the data is shown in Figure 16.2. The plot looks linear over a large portion of its domain, but there is some levelling off at the right end and of course it is basic human biology that we stop growing at some point in the later teens.

The Gompertz curve is often used to describe growth. It can be expressed in the following form:

$$y = \alpha e^{-\beta e^{-\gamma x}}$$

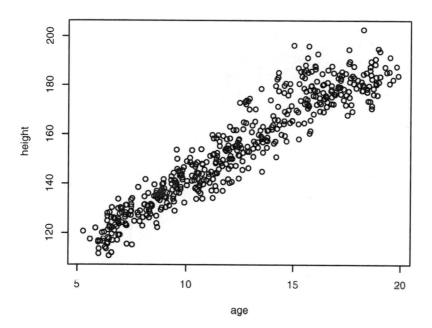

Figure 16.2. Relationship between age and height in `juul2` data set.

The curve has a sigmoid shape, approaching a constant level α as x increases and (in principle) zero for large negative x. The β and γ parameters determine the location and steepness of the transition.

To obtain starting values for a nonlinear fit, one approach is to notice that the relation between y and x is something like log-log linear. Specifically, we can rewrite the relation as

$$\log y = \log \alpha - \beta e^{-\gamma x}$$

which we may rearrange and take logarithms on both sides again, yielding

$$\log(\log \alpha - \log y) = \log \beta - \gamma x$$

That means that if we can come up with a guess for α, then we can guess the two other parameters by a linear fit to transformed data. Since α is the asymptotic maximum, a guess of $\alpha = 200$ could be reasonable. With this guess, we can make a plot that should show an approximately linear relationship ($\log 200 \approx 5.3$):

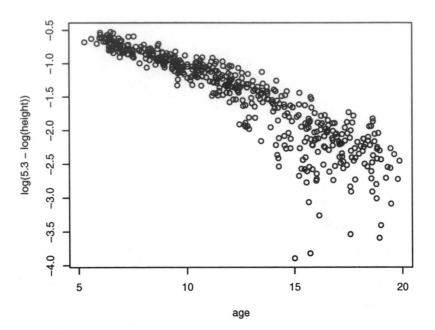

Figure 16.3. Linearized plot of the Gompertz relation when assuming $\alpha \approx 200$.

```
> plot(log(5.3-log(height))~age)
Warning message:
In log(5.3 - log(height)) : NaNs produced
```

Notice that we got a warning that an `NaN` (Not a Number) value was produced. This is because one individual was taller than 200 cm, and we therefore tried to take the logarithm of a negative value. The linearized plot shows a clearly nonconstant variance and probably also some asymmetry of the residual distribution, so the assumptions for linear regression analysis are clearly violated. However, it is good enough for our purpose, and a linear fit gives

```
> lm(log(5.3-log(height))~age)

Call:
lm(formula = log(5.3 - log(height)) ~ age)

Coefficients:
(Intercept)           age
    0.4200       -0.1538

Warning message:
In log(5.3 - log(height)) : NaNs produced
```

Accordingly, an initial guess of the parameters is

$$\log \alpha = 5.3$$
$$\log \beta = 0.42$$
$$\gamma = 0.15$$

Supplying these guesses to nls and fitting the Gompertz curve yields

```
> fit <- nls(height~alpha*exp(-beta*exp(-gamma*age)),
+            start=c(alpha=exp(5.3),beta=exp(0.42),gamma=0.15))
> summary(fit)

Formula: height ~ alpha * exp(-beta * exp(-gamma * age))

Parameters:
        Estimate Std. Error t value Pr(>|t|)
alpha 2.428e+02  1.157e+01  20.978   <2e-16 ***
beta  1.176e+00  1.892e-02  62.149   <2e-16 ***
gamma 7.903e-02  8.569e-03   9.222   <2e-16 ***

Signif. codes:  0 '***' 0.001 '**' 0.01 '*' 0.05 '.' 0.1 ' ' 1

Residual standard error: 6.811 on 499 degrees of freedom

Number of iterations to convergence: 8
Achieved convergence tolerance: 5.283e-06
  (3 observations deleted due to missingness)
```

The final estimates are quite a bit different from the starting values. This reflects the crudeness of the estimation methods used for the starting values. In particular, we used transformations that were based on the mathematical form of the function but did not take the structure of the error variation into account. Also, the important parameter α was obtained by eye.

Looking at the fitted model, however, it is not reassuring that the final estimate for α suggests that boys would continue growing until they are 243 cm tall (for readers in nonmetric countries, that is almost eight feet!). Possibly, the Gompertz curve is just not a good fit for these data.

We can overlay the original data with the fitted curve as follows (Figure 16.4)

```
> plot(age, height)
> newage <- seq(5,20,length=500)
> lines(newage, predict(fit,newdata=data.frame(age=newage)),lwd=2)
```

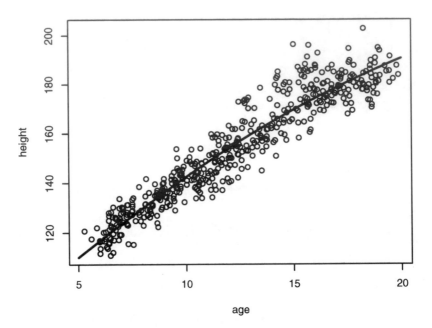

Figure 16.4. The fitted Gompertz curve.

The plot suggests that there is a tendency for the dispersion to increase with increasing fitted values, so we attempt a log-scale fit. This can be done expediently by transforming both sides of the model formula.

```
>
> fit <- nls(log(height)~log(alpha*exp(-beta*exp(-gamma*age)))),
+ start=c(alpha=exp(5.3),beta=exp(.12),gamma=.12))
> summary(fit)

Formula: log(height) ~ log(alpha * exp(-beta * exp(-gamma * age)))

Parameters:
        Estimate Std. Error t value Pr(>|t|)
alpha 255.97694   15.03920  17.021   <2e-16 ***
beta    1.18949    0.02971  40.033   <2e-16 ***
gamma   0.07033    0.00811   8.673   <2e-16 ***
---
Signif. codes:  0 '***' 0.001 '**' 0.01 '*' 0.05 '.' 0.1 ' ' 1

Residual standard error: 0.04307 on 499 degrees of freedom

Number of iterations to convergence: 8
Achieved convergence tolerance: 2.855e-06
   (3 observations deleted due to missingness)
```

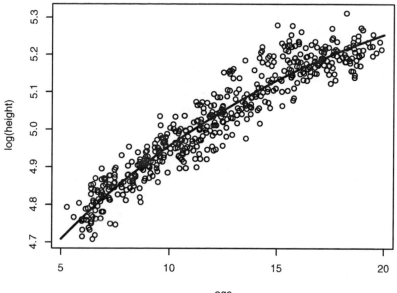

Figure 16.5. Fitted Gompertz curve on log scale.

```
> plot(age, log(height))
> lines(newage, predict(fit,newdata=data.frame(age=newage)),lwd=2)
```

On the log-scale plot (Figure 16.5), the distribution around the curve appears to be more stable. The parameter estimates did not change much, although the maximum height is now increased by a further 13 cm (5 inches) and the γ parameter is reduced to compensate.

Closer inspection of the plots (whether on log scale or not), however, reveals that the Gompertz curve tends to overshoot the data points at the right end, where a much flatter curve would fit the data in the range from 15 years upwards. Although visually there is a nice overall fit, this is not hard to obtain for a three-parameter family of curves, and the Gompertz curves seem unable to fit the characteristic patterns of human growth.

16.3   Self-starting models

Finding starting values is an art rather than a craft, but once a stable method has been found, it may be reasonable to assume that will apply to most data sets from a given model. nls allows the nice feature that the procedure for calculating starting values can be embodied in the expressions that are used on the right-hand side of the model formula. Such functions are by convention named starting with "SS", and R 2.6.2 comes with 10 of these built-in. In particular, there is in fact an SSgompertz function, so we could have saved ourselves much of the trouble of the previous section by just writing

```
> summary(nls(height~SSgompertz(age, Asym, b2, b3)))

Formula: height ~ SSgompertz(age, Asym, b2, b3)

Parameters:
        Estimate Std. Error t value Pr(>|t|)
Asym 2.428e+02  1.157e+01   20.98   <2e-16 ***
b2   1.176e+00  1.892e-02   62.15   <2e-16 ***
b3   9.240e-01  7.918e-03  116.69   <2e-16 ***
...
```

Notice, though, that the parameterization is different: The parameter b3 is actually e^γ, whereas the two other parameters are recognized as α and β.

One minor drawback of self-starting models is that you cannot just transform them if you want to see if the model fits better on, for example, a log scale. In other words, this fails:

```
> nls(log(height) ~ log(SSgompertz(age, Asym, b2, b3)))
Error in nlsModel(formula, mf, start, wts) :
  singular gradient matrix at initial parameter estimates
Calls: nls -> switch -> nlsModel
In addition: Warning message:
In nls(log(height) ~ log(SSgompertz(age, Asym, b2, b3))) :
  No starting values specified for some parameters.
Intializing 'Asym', 'b2', 'b3' to '1.'.
Consider specifying 'start' or using a selfStart model
```

The error message means, in essence, that the self-start machinery is turned off, so nls tries a wild guess, setting all parameters to 1, and then fails to converge from that starting point.

Using expression log(SSgompertz(age, Asym, b2, b3)) to compute the expected value of log(height) is not a problem (in itself). We can take the starting values from the untransformed fit but this is still not enough to make things work.

There is a hitch: SSgompertz returns a *gradient attribute* along with the fitted values. This is the derivative of the fitted value with respect to each of the model parameters. This speeds up the convergence process for the original model but is plainly wrong for the transformed model, where it causes convergence failure. We could patch this up by calculating the correct gradient, but it is expedient simply to discard the attribute by taking as.vector.

```
> cf <- coef(nls(height ~ SSgompertz(age, Asym, b2, b3)))
> summary(nls(log(height) ~
+               log(as.vector(SSgompertz(age,Asym, b2, b3))),
+               start=as.list(cf)))

Formula: log(height) ~ log(as.vector(SSgompertz(age, Asym, b2, b3)))

Parameters:
        Estimate Std. Error t value Pr(>|t|)
Asym  2.560e+02  1.504e+01   17.02   <2e-16 ***
b2    1.189e+00  2.971e-02   40.03   <2e-16 ***
b3    9.321e-01  7.559e-03  123.31   <2e-16 ***
. . .
```

It is possible to write your own self-starting models. It is not hard once you have some experience with R programming, but we shall not go into details here. The essence is that you need two basic items: the model expression and a function that calculates the starting values. You must ensure that these adhere to some formal requirements, and then a constructor function selfStart can be called to create the actual self-starting function.

16.4 Profiling

We discussed profiling before in connection with glm and logistic regression in Section 13.3. For nonlinear regression, there are some slight differences: The function that is being profiled is not the likelihood function but the sum of squared deviations, and the approximate confidence intervals are based on the t distribution rather than the normal distribution. Also, the plotting method does not by default use the signed version of the profile, just the square root of the difference in the sum of squared deviations.

Profiling is designed to eliminate *parameter curvature*. The same model can be formulated using different parameterizations (such as when Gompertz curves could be defined using γ or $b_3 = e^\gamma$). The choice of parameterization can have a substantial influence on whether the distribution of the

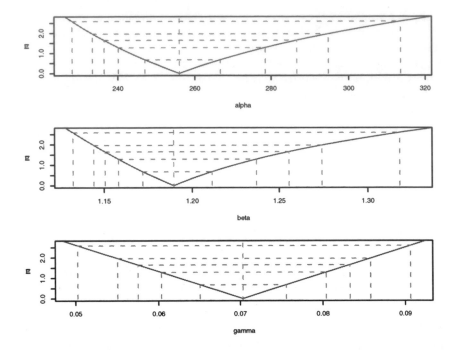

Figure 16.6. Parameter profiles of the Gompertz fit on log scale.

estimate is approximately normal or not, and this in turn means that the use of symmetric confidence intervals based on the standard errors from the model summary can be misleading. Profile-based confidence intervals do not depend on parameterization — if you transform a parameter, the ends of the confidence interval are just transformed in the same way.

There is, however, also *intrinsic curvature* of the models. This describes how far the model is from an approximating linear model. This kind of curvature is independent of parameterization and is harder to adjust for than parameter curvature. The effect of intrinsic curvature is that the *t* distribution used for the calculation of profile-based confidence intervals is not exactly the right distribution to use. Experience suggests that this effect is usually much smaller than the distortions caused by parameter curvature.

For the Gompertz fit (after log transformation), we get the plots shown in Figure 16.6.

```
> par(mfrow=c(3,1))
> plot(profile(fit))
```

The plots show that there is a marked curvature for the α and β parameters, reflected in the curved and asymmetric profiles, whereas the γ profile is more linear and symmetric. This is also seen when comparing the profile-based confidence intervals with those of `confint.default`, which uses the normal approximation and the approximate standard errors.

```
> confint(fit)
Waiting for profiling to be done...
               2.5%          97.5%
alpha 233.49688706 294.76696435
beta    1.14429894   1.27416518
gamma   0.05505754   0.08575007
> confint.default(fit)
               2.5 %         97.5 %
alpha 226.50064512 285.45322721
beta    1.13125578   1.24772846
gamma   0.05443819   0.08622691
```

16.5   Finer control of the fitting algorithm

The Juul example that has been used in this chapter has been quite benign because there are a large number of observations and an objective function that is relatively smooth as a function of the parameters. However, convergence problems easily come up in less nice examples. Nonlinear optimization is simply a tricky topic, to which we have no chance of doing justice in this short chapter. The algorithms have several parameters that can be adjusted in order to help convergence, but since we are not describing the algorithms, it is hardly possible to give more than a feeling for what can be done.

The possibility of supplying a gradient of the fitted curve with respect to parameters was mentioned earlier. If the curve is given by a simple mathematical expression, then the `deriv` function can even be used to generate the gradient automatically. If a gradient is not available, then the algorithm will estimate it numerically; in practice, this often turns out to be equally fast.

The `nls` function features a `trace` argument that, if set to `TRUE`, allows you to follow the parameters and the SSD iteration by iteration. This is sometimes useful to get a handle on what is happening, for instance whether the algorithm is making unreasonably large jumps. To actually modify the behaviour, there is a single `control` argument, which can be set to the return value of `nls.control`, which in turn has arguments to set iteration limits and tolerances (and more).

You can switch out the entire fitting method by using the `algorithm` argument. Apart from the default algorithm, this allows the settings `"plinear"` and `"port"`. The former allows models of the form

$$y = \sum_i \alpha_i f_i(x; \beta_i)$$

that are partially linear since the α_i can be determined by multiple linear regression if the β_i are considered fixed. To specify models with more than one term, you let the expression on the right-hand side of the model formula return a matrix instead of a vector. The latter algorithm uses a routine from the PORT library from Lucent Technologies; this in particular allows you to set contraints on parameters by using the `upper` and `lower` arguments to `nls`.

It should be noted that all the available algorithms operate under the implicit assumption that the SSD(β) is fairly smooth and well behaved, with a well-defined global minimum and no other local minima nearby. There are cases where this assumption is not warranted. In such cases, you might attack the minimization problem directly using the `optim` function.

16.6 Exercises

16.1 Try fitting the Gompertz model for girls in the Juul data. How would you go about testing whether the same model fits both genders?

16.2 The `philion` data contain four small-sample EC50 experiments that are somewhat tricky to handle. We suggest the model $y = y_{max}/(1 + (x/\beta)^\alpha)$. It may be useful to transform y by the square root since the data are counts, and this stabilizes the variance of the Poisson distribution. Consider how to obtain starting values for the model, and fit it with `nls`. The `"port"` algorithm seems more stable for these data. For profiling and confidence intervals, it seems to help if you set the `alphamax` argument to 0.2.

16.3 (Theoretical) Continuing with the `philion` data, consider what happens if you modify the model to be $y = y_{max}/(1 + x/\beta)^\alpha$.

A

Obtaining and installing R and the ISwR package

The way to obtain R is to download it from one of the CRAN (Comprehensive R Archive Network) sites. The main site is

```
http://cran.r-project.org/
```

It has a number of mirror sites worldwide that may be closer to you and give faster download times.

Installation details tend to vary over time, so you should read the accompanying documents and any other information offered on CRAN.

Binary distributions

As of this writing, the version for recent variants of Microsoft Windows comes as a single `R-2.6.2-win32.exe` file, on which you simply double-click with the mouse and then follow the on-screen instructions. When the process is completed, you will have an entry under Programs on the Start menu for invoking R, as well as a desktop icon.

For Linux distributions that use the RPM package format (primarily Red-Hat, Fedora, and SUSE), `.rpm` files of R and the recommended add-on packages can be installed using the `rpm` command and the respective system software management tools. Fedora now has R in its standard repositories, and it is also in the repository of openSUSE.org. Debian packages can be accessed through APT, the Debian package maintenance tool, as can packages for Ubuntu (in both cases, make sure that you get the r-recommended package). Further details are in the FAQ.

For the Macintosh platforms, only OS X 10.2 and above are supported. Installation is by downloading the disk image R-2.6.2.dmg and double-clicking the "R.mpkg" icon found inside it.

Installation from source

Installation of Rfrom source code is possible on all supported platforms, although not quite trivial on Windows, mainly because the build environment is not part of the system. On Unix-like systems (Macintosh OS X included), the process can be as simple as unpacking the sources and writing

```
./configure
make
make install
```

The above works on widely used platforms, provided that the relevant compilers and support libraries are installed. If your system is more esoteric or you want to use special compilers or libraries, then you may need to dig deeper.

For Windows, the directory src/gnuwin32 has an INSTALL file with detailed information about the procedure to follow.

Package installation

To work through the examples and exercises in this book, you should install the ISwR package, which contains the data sets.

Assuming that you are connected to the Internet, you can start R and from the Windows and Macintosh versions use their convenient menu interfaces.

On other platforms, you can type

```
install.packages("ISwR")
```

This will give off a harmless warning and install the package in the default location.

On Unix and Linux systems you will need superuser permissions to install in the system location. Similarly, you may require administrator access on some Windows versions.

Otherwise you can set up a private library directory and install into that. Set the R_LIBS environment variable to use your private library subsequently. Further details can be found on the help page for library.

If your R machine is not connected to the Internet, you can also download the package as a file via a different computer. For Windows and Macintosh, you should get the binary package (.zip or .tgz extension) and then installation from a local file is possible via a menu entry. For Unix and Linux, you can issue the following at the shell prompt (the -l option allows you to give a private library if needed):

```
R CMD INSTALL ISwR
```

More information

Information and further Internet resources for R can be obtained from CRAN and the R homepage at

```
http://www.r-project.org
```

Notice in particular the mailing lists, the user-contributed documents, and the FAQs.

B

Data sets in the I SwR package[1]

I gM	*Immunoglobulin G*

Description

Serum IgM in 298 children aged 6 months to 6 years.

Usage

    IgM

Format

A single numeric vector (g/l).

Source

D.G. Altman (1991), *Practical Statistics for Medical Research*, Table 3.2, Chapman & Hall.

---

[1]Reproduced with permission from the documentation files in the I SwR package.

Examples

```
stripchart(IgM,method="stack")
```

---

alkfos   *Alkaline phosphatase data*

Description

Repeated measurements of alkaline phosphatase in a randomized trial of Tamoxifen treatment of breast cancer patients.

Usage

```
alkfos
```

Format

A data frame with 43 observations on the following 8 variables.

grp a numeric vector, group code (1=placebo, 2=Tamoxifen).
c0 a numeric vector, concentration at baseline.
c3 a numeric vector, concentration after 3 months.
c6 a numeric vector, concentration after 6 months.
c9 a numeric vector, concentration after 9 months.
c12 a numeric vector, concentration after 12 months.
c18 a numeric vector, concentration after 18 months.
c24 a numeric vector, concentration after 24 months.

Source

Original data.

References

B. Kristensen et al. (1994), Tamoxifen and bone metabolism in post-menopausal low-risk breast cancer patients: a randomized study. *Journal of Clinical Oncology*, 12(2):992–997.

ashina	*Ashina's crossover trial*

Description

The `ashina` data frame has 16 rows and 3 columns. It contains data from a crossover trial for the effect of an NO synthase inhibitor on headaches. Visual analog scale recordings of pain levels were made at baseline and at five time points after infusion of the drug or placebo. A score was calculated as the sum of the differences from baseline. Data were recorded during two sessions for each patient. Six patients were given treatment on the first occasion and the placebo on the second. Ten patients had placebo first and then treatment. The order of treatment and the placebo was randomized.

Usage

```
ashina
```

Format

This data frame contains the following columns:

vas.active a numeric vector, summary score when given active substance.

vas.plac a numeric vector, summary score when given placebo treatment.

grp a numeric vector code, 1: placebo first, 2: active first.

Source

Original data.

References

M.Ashina et al. (1999), Effect of inhibition of nitric oxide synthase on chronic tension-type headache: a randomised crossover trial. *Lancet* 353, 287–289

Examples

```
plot(vas.active~vas.plac,pch=grp,data=ashina)
abline(0,1)
```

bcmort	*Breast cancer mortality*

Description

Danish study on the effect of screening for breast cancer.

Usage

```
bcmort
```

Format

A data frame with 24 observations on the following 4 variables.

age a factor with levels 50-54, 55-59, 60-64, 65-69, 70-74, and 75-79.

cohort a factor with levels Study gr., Nat.ctr., Hist.ctr., and Hist.nat.ctr..

bc.deaths a numeric vector, number of breast cancer deaths.

p.yr a numeric vector, person-years under study.

Details

Four cohorts were collected. The "study group" consists of the population of women in the appropriate age range in Copenhagen and Frederiksberg after the introduction of routine mammography screening. The "national control group" consisted of the population in the parts of Denmark in which routine mammography screening was not available. These two groups were both collected in the years 1991–2001. The "historical control group" and the "historical national control group" are similar cohorts from 10 years earlier (1981–1991), before the introduction of screening in Copenhagen and Frederiksberg. The study group comprises the entire population, not just those accepting the invitation to be screened.

Source

A.H. Olsen et al. (2005), Breast cancer mortality in Copenhagen after introduction of mammography screening. *British Medical Journal*, 330: 220–222.

bp.obese	*Obesity and blood pressure*

Description

The bp.obese data frame has 102 rows and 3 columns. It contains data from a random sample of Mexican-American adults in a small California town.

Usage

```
bp.obese
```

Format

This data frame contains the following columns:

sex a numeric vector code, 0: male, 1: female.
obese a numeric vector, ratio of actual weight to ideal weight from New York Metropolitan Life Tables.
bp a numeric vector, systolic blood pressure (mm Hg).

Source

B.W. Brown and M. Hollander (1977), *Statistics: A Biomedical Introduction*, Wiley.

Examples

```
plot(bp~obese,pch = ifelse(sex==1, "F", "M"), data = bp.obese)
```

caesarean	*Caesarean section and maternal shoe size*

Description

The table caesar.shoe contains the relation between caesarean section and maternal shoe size (UK sizes!).

Usage

```
caesar.shoe
```

Format

A matrix with two rows and six columns.

Source

D.G. Altman (1991), *Practical Statistics for Medical Research*, Table 10.1, Chapman & Hall.

Examples

```
prop.trend.test(caesar.shoe["Yes",],margin.table(caesar.shoe,2))
```

---

coking   *Coking data*

Description

The `coking` data frame has 18 rows and 3 columns. It contains the time to coking in an experiment where the oven width and temperature were varied.

Usage

```
coking
```

Format

This data frame contains the following columns:

width a factor with levels 4, 8, and 12, giving the oven width in inches.

temp a factor with levels 1600 and 1900, giving the temperature in Fahrenheit.

time a numeric vector, time to coking.

Source

R.A. Johnson (1994), *Miller and Freund's Probability and Statistics for Engineers*, 5th ed., Prentice-Hall.

Examples

```
attach(coking)
matplot(tapply(time,list(width,temp),mean))
detach(coking)
```

cystfibr	*Cystic fibrosis lung function data*

Description

The cystfibr data frame has 25 rows and 10 columns. It contains lung function data for cystic fibrosis patients (7–23 years old).

Usage

cystfibr

Format

This data frame contains the following columns:

age a numeric vector, age in years.
sex a numeric vector code, 0: male, 1:female.
height a numeric vector, height (cm).
weight a numeric vector, weight (kg).
bmp a numeric vector, body mass (% of normal).
fev1 a numeric vector, forced expiratory volume.
rv a numeric vector, residual volume.
frc a numeric vector, functional residual capacity.
tlc a numeric vector, total lung capacity.
pemax a numeric vector, maximum expiratory pressure.

Source

D.G. Altman (1991), *Practical Statistics for Medical Research*, Table 12.11, Chapman & Hall.

References

O'Neill et al. (1983), The effects of chronic hyperinflation, nutritional status, and posture on respiratory muscle strength in cystic fibrosis, *Am. Rev. Respir. Dis.*, 128:1051–1054.

eba1977	*Lung cancer incidence in four Danish cities 1968–1971*

Description

This data set contains counts of incident lung cancer cases and population size in four neighbouring Danish cities by age group.

Usage

eba1977

Format

A data frame with 24 observations on the following 4 variables:

city a factor with levels Fredericia, Horsens, Kolding, and Vejle.

age a factor with levels 40-54, 55-59, 60-64, 65-69, 70-74, and 75+.

pop a numeric vector, number of inhabitants.

cases a numeric vector, number of lung cancer cases.

Details

These data were "at the center of public interest in Denmark in 1974", according to Erling Andersen's paper. The city of Fredericia has a substantial petrochemical industry in the harbour area.

Source

E.B. Andersen (1977), Multiplicative Poisson models with unequal cell rates, *Scandinavian Journal of Statistics*, 4:153–158.

References

J. Clemmensen et al. (1974), *Ugeskrift for Læger*, pp. 2260–2268.

energy	*Energy expenditure*

Description

The energy data frame has 22 rows and 2 columns. It contains data on the energy expenditure in groups of lean and obese women.

Usage

```
energy
```

Format

This data frame contains the following columns:

expend a numeric vector, 24 hour energy expenditure (MJ).
stature a factor with levels lean and obese.

Source

D.G. Altman (1991), *Practical Statistics for Medical Research*, Table 9.4, Chapman & Hall.

Examples

```
plot(expend~stature,data=energy)
```

ewrates	*Rates of lung and nasal cancer mortality, and total mortality.*

Description

England and Wales mortality rates from lung cancer, nasal cancer, and all causes, 1936–1980. The 1936 rates are repeated as 1931 rates in order to accommodate follow-up for the nickel study.

Usage

```
ewrates
```

Format

A data frame with 150 observations on the following 5 variables:

year calendar period, 1931: 1931–35, 1936: 1936–40, ...
age age class, 10: 10–14, 15:15–19, ...
lung lung cancer mortality rate per 1 million person-years
nasal nasal cancer mortality rate per 1 million person-years
other all cause mortality rate per 1 million person-years

Details

Taken from the "Epi" package by Bendix Carstensen et al.

Source

N.E. Breslow, and N. Day (1987). *Statistical Methods in Cancer Research. Volume II: The Design and Analysis of Cohort Studies*, Appendix IX. IARC Scientific Publications, Lyon.

---

fake.trypsin   *Trypsin by age groups*

Description

The trypsin data frame has 271 rows and 3 columns. Serum levels of immunoreactive trypsin in healthy volunteers (faked!).

Usage

```
fake.trypsin
```

Format

This data frame contains the following columns:

trypsin a numeric vector, serum-trypsin in ng/ml.
grp a numeric vector, age coding. See below.
grpf a factor with levels 1: age 10–19, 2: age 20–29, 3: age 30–39, 4: age 40–49, 5: age 50–59, and 6: age 60–69.

Details

Data have been simulated to match given group means and SD.

Source

D.G. Altman (1991), *Practical Statistics for Medical Research*, Table 9.12, Chapman & Hall.

Examples

```
plot(trypsin~grp, data=fake.trypsin)
```

---

graft.vs.host   *Graft versus host disease*

Description

The gvhd data frame has 37 rows and 7 columns. It contains data from patients receiving a nondepleted allogenic bone marrow transplant with the purpose of finding variables associated with the development of acute graft-versus-host disease.

Usage

```
graft.vs.host
```

Format

This data frame contains the following columns:

pnr a numeric vector patient number.

rcpage a numeric vector, age of recipient (years).

donage a numeric vector, age of donor (years).

type a numeric vector, type of leukaemia coded 1: AML, 2: ALL, 3: CML for acute myeloid, acute lymphatic, and chronic myeloid leukaemia.

preg a numeric vector code indicating whether donor has been pregnant. 0: no, 1: yes.

index a numeric vector giving an index of mixed epidermal cell-lymphocyte reactions.

gvhd a numeric vector code, graft-versus-host disease, 0: no, 1: yes.

time a numeric vector, follow-up time

dead a numeric vector code, 0: no (censored), 1: yes

Source

D.G. Altman (1991), *Practical Statistics for Medical Research*, Exercise 12.3, Chapman & Hall.

Examples

```
plot(jitter(gvhd,0.2)~index,data=graft.vs.host)
```

`heart.rate`   *Heart rates after enalaprilat*

Description

The `heart.rate` data frame has 36 rows and 3 columns. It contains data for nine patients with congestive heart failure before and shortly after administration of enalaprilat, in a balanced two-way layout.

Usage

```
heart.rate
```

Format

This data frame contains the following columns:

hr a numeric vector, heart rate in beats per minute.
subj a factor with levels 1 to 9.
time a factor with levels 0 (before), 30, 60, and 120 (minutes after administration).

Source

D.G. Altman (1991), *Practical Statistics for Medical Research*, Table 12.2, Chapman & Hall.

Examples

```
evalq(interaction.plot(time,subj,hr), heart.rate)
```

`hellung`   *Growth of Tetrahymena cells*

Description

The `hellung` data frame has 51 rows and 3 columns. diameter and concentration of *Tetrahymena* cells with and without glucose added to growth medium.

Usage

```
hellung
```

Format

This data frame contains the following columns:

glucose a numeric vector code, 1: yes, 2: no.
conc a numeric vector, cell concentration (counts/ml).
diameter a numeric vector, cell diameter (μm).

Source

D. Kronborg and L.T. Skovgaard (1990), *Regressionsanalyse*, Table 1.1, FADLs Forlag (in Danish).

Examples

```
plot(diameter~conc,pch=glucose,log="xy",data=hellung)
```

---

intake   *Energy intake*

Description

The intake data frame has 11 rows and 2 columns. It contains paired values of energy intake for 11 women.

Usage

```
intake
```

Format

This data frame contains the following columns:

pre a numeric vector, premenstrual intake (kJ).
post a numeric vector, postmenstrual intake (kJ).

Source

D.G. Altman (1991), *Practical Statistics for Medical Research*, Table 9.3, Chapman & Hall.

Examples

```
plot(intake$pre, intake$post)
```

---

juul	*Juul's IGF data*

Description

The `juul` data frame has 1339 rows and 6 columns. It contains a reference sample of the distribution of insulin-like growth factor (IGF-I), one observation per subject in various ages, with the bulk of the data collected in connection with school physical examinations.

Usage

```
juul
```

Format

This data frame contains the following columns:

age a numeric vector (years).
menarche a numeric vector. Has menarche occurred (code 1: no, 2: yes)?
sex a numeric vector (1: boy, 2: girl).
igf1 a numeric vector, insulin-like growth factor (μg/l).
tanner a numeric vector, codes 1–5: Stages of puberty ad modum Tanner.
testvol a numeric vector, testicular volume (ml).

Source

Original data.

Examples

```
plot(igf1~age, data=juul)
```

---

juul2   *Juul's IGF data, extended version*

Description

The juul2 data frame has 1339 rows and 8 columns; extended version of juul.

Usage

juul2

Format

This data frame contains the following columns:

age a numeric vector (years).

height a numeric vector (cm).

menarche a numeric vector. Has menarche occurred (code 1: no, 2: yes)?

sex a numeric vector (1: boy, 2: girl).

igf1 a numeric vector, insulin-like growth factor (μg/l).

tanner a numeric vector, codes 1–5: Stages of puberty ad modum Tanner.

testvol a numeric vector, testicular volume (ml).

weight a numeric vector, weight (kg).

Source

Original data.

Examples

```
plot(igf1~age, data=juul2)
```

---

kfm   *Breast-feeding data*

Description

The kfm data frame has 50 rows and 7 columns. It was collected by Kim Fleischer Michaelsen and contains data for 50 infants of age approximately 2 months. They were weighed immediately before and after each breast feeding. and the measured intake of breast milk was registered along with various other data.

Usage

```
kfm
```

Format

This data frame contains the following columns:

no a numeric vector, identification number.
dl.milk a numeric vector, breast-milk intake (dl/24h).
sex a factor with levels boy and girl.
weight a numeric vector, weight of child (kg).
ml.suppl a numeric vector, supplementary milk substitute (ml/24h).
mat.weight a numeric vector, weight of mother (kg).
mat.height a numeric vector, height of mother (cm).

Note

The amount of supplementary milk substitute refers to a period before the data collection.

Source

Original data.

Examples

```
plot(dl.milk~mat.height,pch=c(1,2)[sex],data=kfm)
```

---

lung   *Methods for determining lung volume*

Description

The lung data frame has 18 rows and 3 columns. It contains data on three different methods of determining human lung volume.

Usage

```
lung
```

Format

This data frame contains the following columns:

volume a numeric vector, measured lung volume.
method a factor with levels A, B, and C.
subject a factor with levels 1–6.

Source

Anon. (1977), *Exercises in Applied Statistics*, Exercise 4.15, Dept. of Theoretical Statistics, Aarhus University.

---

malaria   *Malaria antibody data*

Description

The malaria data frame has 100 rows and 4 columns.

Usage

malaria

Format

This data frame contains the following columns:

subject subject code.
age age in years.
ab antibody level.
mal a numeric vector code, Malaria: 0: no, 1: yes.

Details

A random sample of 100 children aged 3–15 years from a village in Ghana. The children were followed for a period of 8 months. At the beginning of the study, values of a particular antibody were assessed. Based on observations during the study period, the children were categorized into two groups: individuals with and without symptoms of malaria.

Source

Unpublished data.

Examples

```
summary(malaria)
```

---

melanom	*Survival after malignant melanoma*

Description

The melanom data frame has 205 rows and 7 columns. It contains data relating to the survival of patients after an operation for malignant melanoma, collected at Odense University Hospital by K.T. Drzewiecki.

Usage

```
melanom
```

Format

This data frame contains the following columns:

no a numeric vector, patient code.

status a numeric vector code, survival status; 1: dead from melanoma, 2: alive, 3: dead from other cause.

days a numeric vector, observation time.

ulc a numeric vector code, ulceration; 1: present, 2: absent.

thick a numeric vector, tumor thickness (1/100 mm).

sex a numeric vector code; 1: female, 2: male.

Source

P.K. Andersen, Ø. Borgan, R.D. Gill, and N. Keiding (1991), *Statistical Models Based on Counting Processes*, Appendix 1, Springer-Verlag.

Examples

```
require(survival)
plot(survfit(Surv(days,status==1),data=melanom))
```

nickel	*Nickel smelters in South Wales*

Description

The data concern a cohort of nickel smelting workers in South Wales, with information on exposure, follow-up period, and cause of death.

Usage

nickel

Format

A data frame containing 679 observations of the following 7 variables:

id subject identifier (numeric).
icd ICD cause of death if dead, 0 otherwise (numeric).
exposure exposure index for workplace (numeric)
dob date of birth (numeric).
age1st age at first exposure (numeric).
agein age at start of follow-up (numeric).
ageout age at end of follow-up (numeric).

Details

Taken from the "Epi" package by Bendix Carstensen et al. For comparison purposes, England and Wales mortality rates (per 1,000,000 per annum) from lung cancer (ICDs 162 and 163), nasal cancer (ICD 160), and all causes, by age group and calendar period, are supplied in the data set ewrates.

Source

N.E. Breslow and N. Day (1987). *Statistical Methods in Cancer Research. Volume II: The Design and Analysis of Cohort Studies*, IARC Scientific Publications, Lyon.

---

nickel.expand   *Nickel smelters in South Wales, expanded*

Description

The data concern a cohort of nickel smelting workers in South Wales, with information on exposure, follow-up period, and cause of death, as in the `nickel` data. This version has follow-up times split according to age groups and is merged with the mortality rates in `ewrates`.

Usage

```
nickel.expand
```

Format

A data frame with 3724 observations on the following 12 variables:

agr age class: 10: 10–14, 15: 15–19, ...
ygr calendar period, 1931: 1931–35, 1936: 1936–40, ...
id subject identifier (numeric).
icd ICD cause of death if dead, 0 otherwise (numeric).
exposure exposure index for workplace (numeric).
dob date of birth (numeric).
age1st age at first exposure (numeric).
agein age at start of follow-up (numeric).
ageout age at end of follow-up (numeric).
lung lung cancer mortality rate per 1 million person-years.
nasal nasal cancer mortality rate per 1 million person-years.
other all cause mortality rate per 1 million person-years.

Source

Computed from `nickel` and `ewrates` data sets.

---

philion   *Dose response data*

Description

Four small experiments with the purpose of estimating the EC50 of a biological dose-response relation.

Usage

```
philion
```

Format

A data frame with 30 observations on the following 3 variables:

experiment a numeric vector; codes 1 through 4 denote the experiment number.
dose a numeric vector, the dose.
response a numeric vector, the response (counts).

Details

These data were discussed on the R mailing lists, initially suggesting a log-linear Poisson regression, but actually a relation like $y = y_{max}/(1 + (x/\beta)^{\alpha})$ is more suitable.

Source

Original data from Vincent Philion, IRDA, Québec.

References

```
http://tolstoy.newcastle.edu.au/R/help/03b/1121.html
```

---

react	*Tuberculin reactions*

Description

The numeric vector `react` contains differences between two nurses' determinations of 334 tuberculin reaction sizes.

Usage

```
react
```

Format

A single vector, differences between reaction sizes in mm.

Source

Anon. (1977), *Exercises in Applied Statistics*, Exercise 2.9, Dept. of Theoretical Statistics, Aarhus University.

Examples

```
hist(react) # not good because of discretization effects...
plot(density(react))
```

`red.cell.folate`   *Red cell folate data*

Description

The `folate` data frame has 22 rows and 2 columns. It contains data on red cell folate levels in patients receiving three different methods of ventilation during anesthesia.

Usage

```
red.cell.folate
```

Format

This data frame contains the following columns:

folate a numeric vector, folate concentration (μg/l).

ventilation a factor with levels N2O+O2,24h: 50% nitrous oxide and 50% oxygen, continuously for 24 hours; N2O+O2,op: 50% nitrous oxide and 50% oxygen, only during operation; O2,24h: no nitrous oxide but 35%–50% oxygen for 24 hours.

Source

D.G. Altman (1991), *Practical Statistics for Medical Research*, Table 9.10, Chapman & Hall.

Examples

```
plot(folate~ventilation,data=red.cell.folate)
```

rmr	*Resting metabolic rate*

Description

The rmr data frame has 44 rows and 2 columns. It contains the resting metabolic rate and body weight data for 44 women.

Usage

```
rmr
```

Format

This data frame contains the following columns:

body.weight a numeric vector, body weight (kg).
metabolic.rate a numeric vector, metabolic rate (kcal/24hr).

Source

D.G. Altman (1991), *Practical Statistics for Medical Research*, Exercise 11.2, Chapman & Hall.

Examples

```
plot(metabolic.rate~body.weight,data=rmr)
```

secher	*Birth weight and ultrasonography*

Description

The secher data frame has 107 rows and 4 columns. It contains ultrasonographic measurements of fetuses immediately before birth and their subsequent birth weight.

Usage

```
secher
```

Format

This data frame contains the following columns:

bwt a numeric vector, birth weight (g).
bpd a numeric vector, biparietal diameter (mm).
ad a numeric vector, abdominal diameter (mm).
no a numeric vector, observation number.

Source

D. Kronborg and L.T. Skovgaard (1990), *Regressionsanalyse*, Table 3.1, FADLs Forlag (in Danish).
Secher et al. (1987), European Journal of Obstetrics, Gynecology, and Reproductive Biology, 24: 1–11.

Examples

```
plot(bwt~ad, data=secher, log="xy")
```

---

secretin   *Secretin-induced blood glucose changes*

Description

The `secretin` data frame has 50 rows and 6 columns. It contains data from a glucose response experiment.

Usage

```
secretin
```

Format

This data frame contains the following columns:

gluc a numeric vector, blood glucose level.
person a factor with levels A–E.
time a factor with levels 20, 30, 60, 90 (minutes since injection), and pre (before injection).
repl a factor with levels a: 1st sample; b: 2nd sample.
time20plus a factor with levels 20+: 20 minutes or longer since injection; pre: before injection.
time.comb a factor with levels 20: 20 minutes since injection; 30+: 30 minutes or longer since injection; pre: before injection.

Details

Secretin is a hormone of the duodenal mucous membrane. An extract was administered to five patients with arterial hypertension. Primary registrations (double determination) of blood glucose were on graph paper and later quantified with the smallest of the two measurements recorded first.

Source

Anon. (1977), *Exercises in Applied Statistics*, Exercise 5.8, Dept. of Theoretical Statistics, Aarhus University.

---

stroke   *Estonian stroke data*

Description

All cases of stroke in Tartu, Estonia, during the period 1991–1993, with follow-up until January 1, 1996.

Usage

stroke

Format

A data frame with 829 observations on the following 10 variables.

sex a factor with levels Female and Male.
died a Date, date of death.
dstr a Date, date of stroke.
age a numeric vector, age at stroke.
dgn a factor, diagnosis, with levels ICH (intracranial haemorrhage), ID (unidentified). INF (infarction, ischaemic), SAH (subarchnoid haemorrhage).
coma a factor with levels No and Yes, indicating whether patient was in coma after the stroke.
diab a factor with levels No and Yes, history of diabetes.
minf a factor with levels No and Yes, history of myocardial infarction.
han a factor with levels No and Yes, history of hypertension.
obsmonths a numeric vector, observation times in months (set to 0.1 for patients dying on the same day as the stroke).
dead a logical vector, whether patient died during the study.

Source

Original data.

References

J. Korv, M. Roose, and A.E. Kaasik (1997). Stroke Registry of Tartu, Estonia, from 1991 through 1993. Cerebrovascular Disorders 7:154–162.

`tb.dilute`	*Tuberculin dilution assay*

Description

The `tb.dilute` data frame has 18 rows and 3 columns. It contains data from a drug test involving dilutions of tuberculin.

Usage

```
tb.dilute
```

Format

This data frame contains the following columns:

reaction a numeric vector, reaction sizes (average of diameters) for tuberculin skin pricks.
animal a factor with levels 1–6.
logdose a factor with levels 0.5, 0, and -0.5.

Details

The actual dilutions were 1:100, 1:100$\sqrt{10}$, 1:1000. Setting the middle one to 1 and using base-10 logarithms gives the `logdose` values.

Source

Anon. (1977), *Exercises in Applied Statistics*, part of Exercise 4.15, Dept. of Theoretical Statistics, Aarhus University.

thuesen	*Ventricular shortening velocity*

Description

The thuesen data frame has 24 rows and 2 columns. It contains ventricular shortening velocity and blood glucose for type 1 diabetic patients.

Usage

thuesen

Format

This data frame contains the following columns:

blood.glucose a numeric vector, fasting blood glucose (mmol/l).
short.velocity a numeric vector, mean circumferential shortening velocity (%/s).

Source

D.G. Altman (1991), *Practical Statistics for Medical Research*, Table 11.6, Chapman & Hall.

Examples

```
plot(short.velocity~blood.glucose, data=thuesen)
```

tlc	*Total lung capacity*

Description

The tlc data frame has 32 rows and 4 columns. It contains data on pretransplant total lung capacity (TLC) for recipients of heart-lung transplants by whole-body plethysmography.

Usage

tlc

Format

This data frame contains the following columns:

age a numeric vector, age of recipient (years).
sex a numeric vector code, female: 1, male: 2.
height a numeric vector, height of recipient (cm).
tlc a numeric vector, total lung capacity (l).

Source

D.G. Altman (1991), *Practical Statistics for Medical Research*, Exercise 12.5, 10.1, Chapman & Hall.

Examples

```
plot(tlc~height,data=tlc)
```

---

vitcap   *Vital capacity*

Description

The vitcap data frame has 24 rows and 3 columns. It contains data on vital capacity for workers in the cadmium industry. It is a subset of the vitcap2 data set.

Usage

```
vitcap
```

Format

This data frame contains the following columns:

group a numeric vector; group codes are 1: exposed > 10 years, 3: not exposed.
age a numeric vector, age in years.
vital.capacity a numeric vector, vital capacity (a measure of lung volume) in liters.

Source

P. Armitage and G. Berry (1987), *Statistical Methods in Medical Research*, 2nd ed., Blackwell, p.286.

Examples

```
plot(vital.capacity~age, pch=group, data=vitcap)
```

---

vitcap2   *Vital capacity, full data set*

Description

The vitcap2 data frame has 84 rows and 3 columns. Age and vital capacity for workers in the cadmium industry.

Usage

```
vitcap2
```

Format

This data frame contains the following columns:

group a numeric vector; group codes are 1: exposed > 10 years, 2: exposed < 10 years, 3: not exposed.

age a numeric vector, age in years.

vital.capacity a numeric vector, vital capacity (a measure of lung volume) (l).

Source

P. Armitage and G. Berry (1987), *Statistical Methods in Medical Research*, 2nd ed., Blackwell, p.286.

Examples

```
plot(vital.capacity~age, pch=group, data=vitcap2)
```

---

wright   *Comparison of Wright peak-flow meters*

Description

The wright data frame has 17 rows and 2 columns. It contains data on peak expiratory flow rate with two different flow meters on each of 17 subjects.

Usage

```
wright
```

Format

This data frame contains the following columns:

std.wright a numeric vector, data from large flow meter (l/min).
mini.wright a numeric vector, data from mini flow meter (l/min).

Source

J.M. Bland and D.G. Altman (1986), Statistical methods for assessing agreement between two methods of clinical measurement, *Lancet*, 1:307–310.

Examples

```
plot(wright)
abline(0,1)
```

---

zelazo	*Age at walking*

Description

The zelazo object is a list with four components.

Usage

```
zelazo
```

Format

This is a list containing data on age at walking (in months) for four groups of infants:

active test group receiving active training; these children had their walking and placing reflexes trained during four three-minute sessions that took place every day from their second to their eighth week of life.

passive passive training group; these children received the same types of social and gross motor stimulation, but did not have their specific walking and placing reflexes trained.

none no training; these children had no special training, but were tested along with the children who underwent active or passive training.

ctr.8w eighth-week controls; these children had no training and were only tested at the age of 8 weeks.

Note

When asked to enter these data from a text source, many students will use one vector per group and will need to reformat data into a data frame for some uses. The rather unusual format of this data set mimics that situation.

Source

P.R. Zelazo, N.A. Zelazo, and S. Kolb (1972), "Walking" in the newborn, *Science*, 176: 314–315.

C
Compendium

Elementary

Commands

`ls()` or `objects()`	List objects in workspace
`rm(object)`	Delete `object`
`search()`	Search path

Variable names

Combinations of letters, digits, and period. Must not start with a digit. Avoid starting with period.

Assignments

`<-`	Assign value to variable
`->`	Assignment "to the right"
`<<-`	Global assignment (in functions)

Operators

Arithmetic

+	Addition
−	Subtraction, sign
*	Multiplication
/	Division
^	Raise to power
%/%	Integer division
%%	Remainder from integer division

Logical and relational

==	Equal to
!=	Not equal to
<	Less than
>	Greater than
<=	Less than or equal to
>=	Greater than or equal to
is.na(x)	Missing?
&	Logical AND
|	Logical OR
!	Logical NOT

& and | are *elementwise*. See "Programming" (p. 336) for && and ||.

Vectors and data types

Generating

`numeric(25)`	25 zeros
`character(25)`	25 × `""`
`logical(25)`	25 × FALSE
`seq(-4,4,0.1)`	Sequence: $-4.0, -3.9, 3.8, \ldots, 3.9, 4.0$
`1:10`	Same as `seq(1,10,1)`
`c(5,7,9,13,1:5)`	Concatenation: 5 7 9 13 1 2 3 4 5
`rep(1,10)`	1 1 1 1 1 1 1 1 1 1
`gl(3,2,12)`	Factor with 3 levels, repeat each level in blocks of 2, up to length 12 (i.e., 1 1 2 2 3 3 1 1 2 2 3 3)

Coercion

`as.numeric(x)`	Convert to numeric
`as.character(x)`	Convert to text string
`as.logical(x)`	Convert to logical
`factor(x)`	Create factor from vector x

For factors, see also "Tabulation, grouping, and recoding" (p. 331).

Data frames

`data.frame(height = c(165,185), weight = c(90,65))`	Data frame with two named vectors
`data.frame(height, weight)`	Collect vectors into data frame
`dfr$var`	Select vector var in data frame dfr
`attach(dfr)`	Put data frame in search path
`detach()`	— and remove it from path

Attached data frames always come *after* `.GlobalEnv` in the search path.

Attached data frames are copies; subsequent changes to dfr have no effect.

Numerical functions

Mathematical

log(x)	Logarithm of x, natural (base-e) logarithm
log10(x)	Base-10 logarithm
exp(x)	Exponential function e^x
sin(x)	Sine
cos(x)	Cosine
tan(x)	Tangent
asin(x)	Arcsin (inverse sine)
acos(x)	
atan(x)	
min(x)	Smallest value in vector
min(x1,x2,...)	Minimum over several vectors (one number)
max(x)	Largest value in vector
range(x)	Like c(min(x),max(x))
pmin(x1,x2,...)	Parallel (elementwise) minimum over multiple equally long vectors
pmax(x1,x2,...)	Parallel maximum
length(x)	Number of elements in vector
sum(complete.cases(x))	Number of nonmissing elements in vector

Statistical

mean(x)	Average
sd(x)	Standard deviation
var(x)	Variance
median(x)	Median
quantile(x,p)	Quantiles
cor(x,y)	Correlation

Indexing/selection

`x[1]`	First element
`x[1:5]`	Subvector containing first five elements
`x[c(2,3,5,7,11)]`	Element nos. 2, 3, 5, 7, and 11
`x[y<=30]`	Selection by logical expression
`x[sex=="male"]`	Selection by factor variable
`i <- c(2,3,5,7,11); x[i]`	Selection by numeric variable
`l <- (y<=30); x[l]`	Selection by logical variable

Matrices and data frames

`m[4,]`	Fourth row
`m[,3]`	Third column
`dfr[dfr$var<=30,]`	Partial data frame
`subset(dfr,var<=30)`	Same, often simpler

Input of data

`data(name)`	Built-in data set
`read.table("filename")`	Read from external file

Common arguments to `read.table`

`header=TRUE`	First line has variable names
`sep=","`	Data are separated by commas
`dec=","`	Decimal point is comma
`na.strings="."`	Missing value is dot

Variants of `read.table`

`read.csv("filename")`	Comma-separated
`read.delim("filename")`	Tab-delimited
`read.csv2("filename")`	Semicolon-separated, comma decimal point
`read.delim2("filename")`	Tab-delimited, comma decimal point

These all set `header=TRUE`.

Missing values

Functions

`is.na(x)`	Logical vector. TRUE where x has NA.
`complete.cases(x1,x2,...)`	Missing neither in x1, nor x2, nor....

Arguments to other functions

`na.rm=`	In statistical functions, remove missing if TRUE, return NA if FALSE.
`na.last=`	In `sort`; TRUE, FALSE and NA mean, respectively, "last", "first", and "throw away".
`na.action=`	In `lm`, etc., values `na.fail`, `na.omit`, `na.exclude`; also in `options("na.action")`.
`na.print=`	In `summary` and `print.default`; how to represent NA in output.
`na.strings=`	In `read.table()`; code(s) for NA in input.

Tabulation, grouping, and recoding

`table(f1,...)`	(Cross)-tabulation
`xtabs(~ f1 + ...)`	ditto, formula interface
`ftable(f1 ~ f2 + ...)`	"Flat" tables
`tapply(x,f,mean)`	Table of means
`aggregate(df,list(f),mean)`	Means for several variables
`by(df, list(f), summary)`	Summarize data frame by group
`factor(x)`	Convert vector to factor
`cut(x,breaks)`	Groups from cutpoints for continuous variable

Arguments to `factor`

`levels`	Values of x to code. Use if some values are not present in data or if the order would be wrong.
`labels`	Values associated with factor levels.
`exclude`	Values to exclude. Default NA. Set to NULL to have missing values included as a level.

Arguments to `cut`

`breaks`	Cutpoints. Note that values of x outside `breaks` give NA.
`labels`	Names for groups. Default is `(0,30]`, etc.
`right`	Right endpoint included? (FALSE: left)

Recoding factors

`levels(f) <- names`	New level names

`levels(f) <- list(`
 `new1=c("old1","old2")`
 `new2="old3")`
	Combining levels

Statistical distributions

Normal distribution

`dnorm(x)`	Density
`pnorm(x)`	Cumulative distribution function, $P(X \leq x)$
`qnorm(p)`	p-quantile, $x : P(X \leq x) = p$
`rnorm(n)`	n (pseudo-)random normally distributed numbers

Distributions

`pnorm(x,mean,sd)`	Normal
`plnorm(x,mean,sd)`	Lognormal
`pt(x,df)`	Student's t
`pf(x,n1,n2)`	F distribution
`pchisq(x,df)`	χ^2
`pbinom(x,n,p)`	Binomial
`ppois(x,lambda)`	Poisson
`punif(x,min,max)`	Uniform
`pexp(x,rate)`	Exponential
`pgamma(x,shape,scale)`	Gamma
`pbeta(x,a,b)`	Beta

Same convention (d-q-r) for density, quantiles, and random numbers as for normal distribution.

Statistical standard methods

Continuous response

`t.test`	One- and two-sample t tests
`pairwise.t.test`	Pairwise comparisons
`cor.test`	Correlation
`var.test`	Comparison of two variances (F test)
`lm(y ~ x)`	Regression analysis
`lm(y ~ f)`	One-way analysis of variance
`lm(y ~ f1 + f2)`	Two-way analysis of variance
`lm(y ~ f + x)`	Analysis of covariance
`lm(y ~ x1 + x2 + x3)`	Multiple regression analysis
`bartlett.test`	Bartlett's test (k variances)
Nonparametric:	
`wilcox.test`	One- and two-sample Wilcoxon tests
`kruskal.test`	Kruskal–Wallis test
`friedman.test`	Friedman's two-way analysis of variance
`cor.test` variants:	
`method="kendall"`	Kendall's τ
`method="spearman"`	Spearman's ρ

Discrete response

`binom.test`	Binomial test (incl. sign test)
`prop.test`	Comparison of proportions
`prop.trend.test`	Test for trend in relative proportions
`fisher.test`	Exact test in small tables
`chisq.test`	χ^2 test
`glm(y ~ x1+x2+x3, binomial)`	Logistic regression

Models

Model formulas

~	Described by
+	Additive effects
:	Interaction
*	Main effects + interaction
	(a*b = a + b + a:b)
-1	Remove intercept

Classifications are represented by descriptive variable being a *factor*.

Linear, nonlinear, and generalized linear models

`lm.out <- lm(y ~ x)`	Fit model and save result
`summary(lm.out)`	Coefficients, etc.
`anova(lm.out)`	Analysis of variance table
`fitted(lm.out)`	Fitted values
`resid(lm.out)`	Residuals
`predict(lm.out, newdata)`	Predictions for new data frame
`glm(y ~ x, binomial)`	Logistic regression
`glm(y ~ x, poisson)`	Poisson regression
`nls(y ~ a*exp(-b*x), start=c(a=5, b=.2))`	Nonlinear regression

Diagnostics

`rstudent(lm.out)`	Studentized residuals
`dfbetas(lm.out)`	Change in β if obs. removed
`dffits(lm.out)`	Change in fit if obs. removed

Survival analysis

`S <- Surv(time, ev)`	Create survival object
`survfit(S)`	Kaplan–Meier estimate
`plot(survfit(S))`	Survival curve
`survdiff(S ~ g)`	(Log-rank) test for equal survival curves
`coxph(S ~ x1 + x2)`	Cox's proportional hazards model

Graphics

Standard plots

plot()	Scatterplot (and more)
hist()	Histogram
boxplot()	Box-and-whiskers plot
stripplot()	Stripplot
barplot()	Bar diagram
dotplot()	Dot diagram
piechart()	Cakes...
interaction.plot()	Interaction plot

Plotting elements

lines()	Lines
abline()	Line given by intercept and slope (and more)
points()	Points
segments()	Line segments
arrows()	Arrows (N.B.: angle=90 for error bars)
axis()	Axis
box()	Frame around plot
title()	Title (above plot)
text()	Text in plot
mtext()	Text in margin
legend()	List of symbols

These are all *added* to existing plots.

Graphical parameters

pch	Symbol (*plotting character*)
mfrow, mfcol	Several plots on one (*multi*frame)
xlim, ylim	Plot limits
lty, lwd	Line type/width
col	Colour
cex, mex	Character size and line spacing in margins

See the help page for par for more details.

Programming

| Conditional execution | ```
if(p<0.05)
    print("Hooray!")
``` |
|---|---|
| — with alternative | ```
if(p<0.05)
    print("Hooray!")
else
    print("Bah.")
``` |
| Loop over list | ```
for(i in 1:10)
    print(i)
``` |
| Loop | ```
i <- 1
while(i<10) {
    print(i)
    i <- i + 1
}
``` |
| User-defined function | ```
f <- function(a,b,doit=FALSE){
    if (doit)
        a + b
    else
        0
}
``` |

In flow control, one uses a `&&` b and a `||` b, where b is only computed if necessary; that is, `if a then b else FALSE` and `if a then TRUE else b`.
```

D

Answers to exercises

1.1 One possibility is

```
x <- y <- c(7, 9, NA, NA, 13)
all(is.na(x) == is.na(y)) & all((x == y)[!is.na(x)])
```

Notice that FALSE & NA is FALSE, so the case of different NA patterns is handled correctly.

1.2 Factor x gets treated as if it contained the integer codes.

```
x <- factor(c("Huey", "Dewey", "Louie", "Huey"))
y <- c("blue", "red", "green")
x
y[x]
```

(This is useful, e.g., when selecting plot symbols.)

1.3

```
juul.girl <- juul[juul$age >=7 & juul$age < 14 & juul$sex == 2,]
summary(juul.girl)
```

1.4 The levels with the same name are collapsed into one.

1.5 `sapply(1:10, function(i) mean(rexp(20)))`

2.1 To insert 1.23 between x[7] and x[8]:

```
x <- 1:10
z <- append(x, 1.23, after=7)
z
```

Otherwise, consider

```
z <- c(x[1:7],1.23,x[8:10])
z
```

or, more generally, to insert v just after index k (the boundary cases require some care),

```
v <- 1.23; k <- 7
i <- seq(along=x)
z <- c(x[i <= k], v, x[i > k])
z
```

2.2   (First part only) Use

```
write.table(thuesen, file="foo.txt")
# edit the file
read.table("foo.txt", na.strings=".")
```

or

```
write.table(thuesen, file="foo.txt", na=".")
read.table("foo.txt", na.strings=".")
```

(Notice that if you do not edit the file in the first case, then the second column gets read as a character vector.)

3.1

```
1 - pnorm(3)
1 - pnorm(42, mean=35, sd=6)
dbinom(10, size=10, prob=0.8)
punif(0.9) # this one is obvious...
1 - pchisq(6.5, df=2)
```

It might be better to use `lower.tail=FALSE` instead of subtracting from 1 in (a), (b), and (e). Notice that question (c) is about a point probability, whereas the others involve the cumulative distribution function.

3.2   Evaluate each of the following. Notice that the standard normal can be used for all questions.

```
pnorm(-2) * 2
qnorm(1-.01/2)
qnorm(1-.005/2)
qnorm(1-.001/2)
qnorm(.25)
qnorm(.75)
```

Again, lower.tail can be used in some cases.

3.3 dbinom(0, size=10, prob=.2)

3.4   Either of the following should work:

```
rbinom(10, 1, .5)
ifelse(rbinom(10, 1, .5) == 1, "H", "T")
c("H", "T")[1 + rbinom(10, 1, .5)]
```

The first one gives a 0/1 result, the two others H/T like the sample example in the text. One advantage of using rbinom is that its prob argument
can be a vector, so you can have different probabilities of success for each
element of the result.

4.1   For example,

```
x <- 1:5 ; y <- rexp(5,1) ; opar <- par(mfrow=c(2,2))
plot(x, y, pch=15) # filled square
plot(x, y, type="b", lty="dotted")
plot(x, y, type="b", lwd=3)
plot(x, y, type="o", col="blue")
par(opar)
```

4.2   Use a filled symbol, and set the fill colour equal to the plot
background:

```
plot(rnorm(10),type-"o", pch=21, bg="white")
```

4.3   You can use qqnorm with plot.it=F and get a return value from
which you can extract the range information (you could of course also get
this "by eye").

```
x1 <- rnorm(20)
x2 <- rnorm(10)+1
q1 <- qqnorm(x1, plot.it=F)
q2 <- qqnorm(x2, plot.it=F)
xr <- range(q1$x, q2$x)
yr <- range(q1$y, q2$y)
qqnorm(x1, xlim=xr, ylim=yr)
points(q2, col="red")
```

Here, qqnorm is used for the basic plot to get the labels right. Then
points is used with q2 for the overlay.

Setting type="l" gives a messy plot because the values are not plotted
in order. The remedy is to use sort(x1) and sort(x2).

4.4   The breaks occur at integer values, as do the data. Data on the boundary are counted in the column to the left of it, effectively shifting the

histogram half a unit left. The truehist function allows you to specify a
better set of breaks.

```
hist(react)
library(MASS)
truehist(react,h=1,x0=.5)
```

4.5   The thing to notice is the linear interpolation between data points:

```
z <- runif(5)
curve(quantile(z,x),  from=0,  to=1)
```

5.1   The distribution appears reasonably normal, with some discretization effect and two weak outliers, one at each end. There is a significant
difference from zero ($t = -7.75, p = 1.1 \times 10^{-13}$).

```
qqnorm(react)
t.test(react)
```

5.2   t.test(vital.capacity~group,conf=0.99,data=vitcap).
The fact that age also differs by group may cause bias.

5.3   This is quite parallel to t.test usage

```
wilcox.test(react)
wilcox.test(vital.capacity~group, data=vitcap)
```

5.4   The following builds a post-vs.-pre plot, a difference-vs.-average)
(Bland-Altman) plot, and a histogram and a Q-Q plot of the differences.

```
attach(intake) ; opar <- par(mfrow=c(2,2))
plot(post ~ pre) ; abline(0,1)
plot((post+pre)/2, post - pre,
     ylim=range(0,post-pre)); abline(h=0)
hist(post-pre)
qqnorm(post-pre)
detach(intake)
par(opar)
```

5.5   The outliers are the first and last observations in the (sorted) data
vector and can be removed as follows

```
shapiro.test(react)
shapiro.test(react[-c(1,334)])
qqnorm(react[-c(1,334)])
```

The test comes out highly significant even with outliers removed because
it picks up the discretization effect in the otherwise nearly straight-line
qqnorm plot.

5.6 A paired *t* test is appropriate if there is no period effect. However, even with a period effect (assumed additive), you would expect the difference between the two periods to be the same in both groups if there were no effect of treatment. This can be used to test for a treatment effect.

```
attach(ashina)
t.test(vas.active, vas.plac, paired=TRUE)
t.test((vas.active-vas.plac)[grp==1],
       (vas.plac-vas.active)[grp==2])
```

Notice that the subtraction is reversed in one group. Observe that the confidence interval in the second case is for *twice* the treatment effect.

5.7 This is the sort of thing `replicate` is for. The plot at the end shows a P-P plot with logarithmic axes, showing that extreme *p*-values tend to be exaggerated.

```
t.test(rnorm(25))$p.value        #repeat 10x
t.test(rt(25,df=2))$p.value      #repeat 10x
t.test(rexp(25), mu=1)$p.value   #repeat 10x
x <- replicate(5000, t.test(rexp(25), mu=1)$p.value)
qqplot(sort(x),ppoints(5000),type="l",log="xy")
```

6.1 The following gives both elementary and more general answers. Notice the use of `confint`.

```
fit <- lm(metabolic.rate ~ body.weight, data=rmr)
summary(fit)
811.2267 + 7.0595 * 70 # , or:
predict(fit, newdata=data.frame(body.weight=70))
qt(.975,42)
7.0595 + c(-1,1) * 2.018 * 0.9776 # , or:
confint(fit)
```

6.2 `summary(lm(sqrt(igf1)~age,data=juul,subset=age>25))`

6.3 We can fit a linear model and plot the data as follows:

```
summary(lm(log(ab)~age, data=malaria))
plot(log(ab)~age, data=malaria)
```

The plot appears to show a cyclic pattern. It is unclear whether it reflects a significant departure from the model, though. Malaria is a disease with epidemic behaviour, so cycles are plausible.

6.4 (This could be elaborated by wrapping the random number generation in a function, etc.)

```
rho <- .90 ; n <- 100
x <- rnorm(n)
```

```
y <- rnorm(n, rho * x, sqrt(1 - rho^2))
plot(x, y)
cor.test(x, y)
cor.test(x, y, method="spearman")
cor.test(x, y, method="kendall")
```

You will most likely find that the Kendall correlation is somewhat smaller than the two others.

7.1

```
walk <- unlist(zelazo) # or c(..,recursive=TRUE)
group <- factor(rep(1:4,c(6,6,6,5)), labels=names(zelazo))
summary(lm(walk ~ group))
t.test(zelazo$active,zelazo$ctr.8w) # first vs. last
t.test(zelazo$active,unlist(zelazo[-1])) # first vs. rest
```

7.2 A and C differ with B intermediate, not significantly different from either. (The B–C comparison is not available from the summary, but due to the balanced design, the standard error of that difference is 0.16656 like the two others.)

```
fit <- lm(volume~method+subject, data=lung)
anova(fit)
summary(fit)
```

7.3

```
kruskal.test(walk ~ group)
wilcox.test(zelazo$active,zelazo$ctr.8w) # first vs. last
wilcox.test(zelazo$active,unlist(zelazo[-1])) # first vs. rest
friedman.test(volume ~ method | subject, data=lung)
wilcox.test(lung$volume[lung$method=="A"],
            lung$volume[lung$method=="C"], paired=TRUE) # etc.
```

7.4 (Only the square-root transform is shown; you can do the same for log-transformed and untransformed data.)

```
attach(juul)
tapply(sqrt(igf1),tanner, sd, na.rm=TRUE)
plot(sqrt(igf1)~jitter(tanner))
oneway.test(sqrt(igf1)~tanner)
```

The square root looks nice, logarithms become skewed in the opposite direction. The transformations do not make much of a difference for the test. It is, however, a problem that strong age effects are being ignored, particularly within Tanner stage 1.

8.1 With 10 patients, $p = 0.1074$. Fourteen or more are needed for significance at level 0.05.

```
binom.test(0, 10, p=.20, alt="less")
binom.test(0, 13, p=.20, alt="less")
binom.test(0, 14, p=.20, alt="less")
```

8.2   Yes, it is highly significant.

```
prop.test(c(210,122),c(747,661))
```

8.3   The confidence interval (from `prop.test`) is $(-0.085, 0.507)$

```
M <- matrix(c(23,7,18,13),2,2)
chisq.test(M)
fisher.test(M)
prop.test(M)
```

8.4   The following is a simplified analysis, which uses `fisher.test` because of the small cell counts:

```
tbl <- c(42, 157, 47, 62, 4, 15, 4, 1, 8, 28, 9, 7)
dim(tbl) <- c(2,2,3)
dimnames(tbl) <- list(c("A","B"),
                      c("not pierced","pierced"),
                      c("ok","broken","cracked"))
ftable(tbl)
fisher.test(tbl["B",,]) # slice analysis
fisher.test(tbl["A",,])
fisher.test(margin.table(tbl,2:3)) # marginal
```

You may wish to check that there is little or no effect of egg size on breakage, so that the marginal analysis is defensible. You could also try collapsing the "broken" and "cracked" categories.

8.5   The curve shows substantial discontinuities where probability mass is shifted from one tail to the other and also a number of local minima. A confidence region could be defined as those p against which there is no significant evidence at level α, but for some α that is not an interval.

```
p <- seq(0,1,0.001)
pval <- sapply(p,function(p)binom.test(3,15,p=p)$p.value)
plot(p,pval,type="l")
```

9.1   The estimated sample size is 6.29 or 8.06 per group depending on whether you use one- or two-sided testing. The approximate formula gives 6.98 for the two-sided case. The reduction in power due to the un-balanced sampling can be accounted for by reducing `delta` by the ratio of the two SEDM.

```
power.t.test(power=.8,delta=.30,sd=.20)
power.t.test(power=.8,delta=.30,sd=.20,alt="one.sided")
```

```
(qnorm(.975)+qnorm(.8))^2*2*(.2/.3)^2 # approx. formula
power.t.test(n=8, delta=.30, sd=.20)    # power with eq.size
d2 <- .30 * sqrt(2/8) / sqrt(1/6+1/10) # corr.f.uneq. size
power.t.test(n=8, delta=d2, sd=.20)
```

9.2   This is straightforward:

```
power.prop.test(power=.9, p1=.6, p2=.75)
power.prop.test(power=.8, p1=.6, p2=.75)
```

9.3   Notice that the noncentral *t* distribution is asymmetric, with a rather heavy right tail.

```
curve(dt(x-3, 25), from=0, to=5)
curve(dt(x, 25, 3), add=TRUE)
```

9.4   This causes the "power" at zero effect size (i.e., under the null hypothesis) to be *half* the significance level, in contradiction to theory. For any relevant true effect size, the difference is immaterial.

9.5   The power in that case is approximately 0.50; exactly so if the variance is assumed known.

10.1

```
attach(thuesen)
f <- cut(blood.glucose, c(4, 7, 9, 12, 20))
levels(f) <- c("low", "intermediate", "high", "very high")
```

10.2

```
bcmort2 <- within(bcmort, {
  period <- area <- cohort
  levels(period) <- rep(c("1991-2001","1981-1991"), each=2)
  levels(area) <- rep(c("Cph+Frb","Nat"),2)
})
summary(bcmort2)
```

10.3   One way is the following (for later use, we also make sure that variables are converted to factors):

```
ashina.long <- reshape(ashina, direction="long",
                       varying=1:2, timevar="treat")
ashina.long <- within(ashina.long, {
    m <- matrix(c(2,1,1,2),2)
    id <- factor(id)
    treat <- factor(treat)
    grp <- factor(grp)
    period <- factor(m[cbind(grp,treat)])
```

```
    rm(m)
})
```

Notice the use of *array indexing*. Alternatively, an `ifelse` construct can be used; e.g., the following (notice that `(3 - grp)` is 2 when `grp` is 1 and vice versa):

```
within(ashina.long,
  period2 <- ifelse(treat != "active",
              as.numeric(grp), 3 - as.numeric(grp))
)
```

Arithmetic involving `grp` does not work after it was converted to a factor, hence the conversion with `as.numeric`.

10.4 This can be done a little more easily than in the `nickel` example by using `subset` and `transform`. It also helps that all observation periods start at time zero in this case.

```
stroke.trim <- function(t1, t2)
   subset(transform(stroke,
                    entry=t1, exit=pmin(t2, obsmonths),
                    dead=dead & obsmonths <= t2),
          entry < exit)
stroke2 <- do.call(rbind, mapply(stroke.trim,
        c(0,0.5,2,12), c(0.5,2,12,Inf), SIMPLIFY=F))
table(stroke$dead)
table(stroke2$dead)
```

Notice the use of `mapply` here. This is like `sapply` and `lapply` but allows the function to have multiple arguments. Alternatively, one could arrange for `stroke.trim` to have a single `interval` argument and use `lapply` on a list of such intervals.

The tabulation at the end is a "sanity check" to show that we have the same number of deaths but many more censored cases after time-splitting.

11.1 The model with both diameters has a residual error of 0.107, compared with 0.128 using abdominal diameter alone and 0.281 with no predictors at all. If a fetus is scaled isotropically, a cubic relation with weight is expected, and you could speculate that this is reflected in the sum of coefficients when using log scales.

```
summary(lm(log(bwt) ~ log(bpd) + log(ad), data=secher))
summary(lm(log(bwt) ~ log(ad), data=secher))
```

11.2 If you use `attach(tlc)`, the `tlc` variable will mask the data frame of the same name, which makes it awkward to access the data frame if you need to. If the data frame is in the global environment rather than in a package, you get the opposite problem, masking of the variable by the data frame. The simplest workaround is to avoid `attach`.

```
pairs(tlc)
summary(lm(log(tlc) ~ ., data=tlc))
opar <- par(mfrow=c(2,2))
plot(lm(log(tlc) ~ ., data=tlc), which=1:4)

drop1(lm(log(tlc) ~ ., data=tlc))
drop1(lm(log(tlc) ~ . - age, data=tlc))

par(mfrow=c(1,1))
plot(log(tlc) ~ height, data=tlc)
par(mfrow=c(2,2))
plot(lm(tlc ~ ., data=tlc), which=1:4) # slightly worse
par(opar)
```

Some new variations of model formulas were introduced above. A dot on the right-hand side in this context means "everything not used on the left-hand side" within the scope of the data frame. A minus term is removed from the model. In other words, ... ~ . - age is the same as ... ~ sex + height.

11.3   The regression coefficient describes a value to be added for females.

11.4   age is highly significant in the first analysis but only borderline significant ($p = 0.06$) in the second analysis after removing height and weight. You would expect similar results, but the number of observations differs in the two cases, due to missing observations.

```
summary(lm(sqrt(igf1) ~ age, data=juul2, subset=(age >= 25)))
anova(lm(sqrt(igf1) ~ age + weight + height,
         data=juul2, subset=(age >= 25)))
```

11.5   sex is treated as a binary indicator for girls. Notice that there are effects both of the mother's and the child's size. The reason why height rather than weight of the mother enters into the equation is somewhat obscure, but one could speculate that weight is an unreliable indicator shortly after pregnancy.

```
summary(lm(dl.milk ~ . - no, data=kfm))
summary(lm(dl.milk ~ . - no - mat.weight, data=kfm))
summary(lm(dl.milk ~ . - no - mat.weight - sex, data=kfm))
summary(lm(dl.milk ~ weight + mat.height, data=kfm))
```

The variations on model formulas used here were described in the solution to Exercise 11.2.

12.1   Using `ashina.long` from Exercise 10.3,

```
fit.ashina <- lm(vas ~ id + period + treat, data=ashina.long)
drop1(fit.ashina, test="F")
anova(fit.ashina)

attach(ashina)
dd <- vas.active - vas.plac
t.test(dd[grp==1], -dd[grp==2], var.eq=T)
t.test(dd[grp==1], dd[grp==2], var.eq=T)
```

Notice that the imbalance in group sizes makes the tests for period and treatment effects order-dependent. The *t* tests are equivalent to the *F* tests from `drop1` but not those from `anova`.

12.2

```
attach(tb.dilute)
anova(lm(reaction ~ animal + logdose))
ld <- c(0.5, 0, -0.5)[logdose]
anova(lm(reaction ~ animal + ld))
summary(lm(reaction ~ animal + ld))
4.7917 + 0.6039 * qt(c(.025,.975), 11)
# or:
confint(lm(reaction ~ animal + ld))["ld",]

slopes <- reaction[logdose==0.5] - reaction[logdose==-0.5]
t.test(slopes)

anova(lm(reaction ~ animal*ld))
```

Notice that the formula for the fitted slope is $\hat{\beta} = \sum xy / \sum x^2$ since $x = 0$, which reduces to taking differences. (The calculation does rely on data being in the right order.)

The confidence interval is wider in the *t* test, reflecting that slopes may vary between rats and that there are fewer degrees of freedom for estimating the variation.

The final ANOVA contains a test for parallel slopes, and the *F* statistic is less than one, so in these data the slopes vary *less* than expected and the DF must be the important issue for the confidence interval.

12.3   This can be varied indefinitely, but consider these examples:

```
model.matrix(~ a:b)          ; lm(z ~ a:b)
model.matrix(~ a * b)        ; lm(z ~ a * b)
model.matrix(~ a:x)          ; lm(z ~ a:x)
model.matrix(~ a * x)        ; lm(z ~ a * x)
model.matrix(~ b * (x + y))  ; lm(z ~ b * (x + y))
```

The first model is singular because indicator variables are created for all four groups, but the intercept is not removed. R will only reduce the set of design variables for an interaction term between categorical variables when one of the main effects is present. There are no singularities in either of the two cases involving a categorical and a continuous variable, but the first one has one parameter less (common-intercept model).

The last example has a "coincidental" singularity (x and y are proportional within each level of b) that R has no chance of detecting.

12.4 The models can be illustrated by plotting the fitted values against time with separate symbols for each person; e.g.,

```
tt <- c(20,30,60,90,0)[time]
plot(fitted(model4)~tt,pch=as.character(person))
```

With model1 there is no imposed structure, model2 is completely additive so that the individual traces are parallel to each other, model3 allows the jump from the "pre" value to the value at 20 minutes to vary between individuals, and finally model4 is like model3 except that there is no change after 30 minutes (traces become horizontal). So model3 is nested in model1 and both model2 and model4 are nested in model3, but there is no nesting relation between model2 and model4.

12.5

```
bp.obese <- transform(bp.obese,sex=factor(sex, labels=c("M","F")))
plot(log(bp) ~ log(obese), pch=c(20,21)[sex], data=bp.obese)
summary(lm(log(bp) ~ sex, data=bp.obese))
summary(lm(log(bp) ~ sex + log(obese), data=bp.obese))
summary(lm(log(bp) ~ sex*log(obese), data=bp.obese))
```

12.6

```
vitcap2 <- transform(vitcap2,group=factor(group,
                                          labels=c("exp>10",
                                          "exp<10", "unexp")))
attach(vitcap2)
plot(vital.capacity~age, pch=(20:22)[group])
vit.fit <- lm(vital.capacity ~ age*group)
summary(vit.fit)
drop1(vit.fit, test="F")
for (i in 1:3) abline(lm(vital.capacity ~ age,
                     subset=as.numeric(group)==i), lty=i)
legend(20, 3.5 ,legend=levels(group), pch=20:22, lty=1:3)
```

Notice that there is a significant interaction; i.e., the lines are *not* parallel.

12.7

```
juul.prepub <- subset(juul, tanner==1)

summary(lm(sqrt(igf1)~age, data=juul.prepub, subset= sex==1))
summary(lm(sqrt(igf1)~age, data=juul.prepub, subset= sex==2))
```

```
summary(lm(sqrt(igf1)~age*factor(sex), data=juul.prepub))
summary(lm(sqrt(igf1)~age+factor(sex), data=juul.prepub))
```

12.8

```
summary(fit.aicopt <- step(lm(dl.milk ~ . - no, data=kfm)))
opar <- par(mfrow=c(2,2))
plot(fit.aicopt, which=1:4)
kfm[32,]
summary(kfm)
summary(update(fit.aicopt, ~ . - sex))
plot(update(fit.aicopt, ~ . - sex - ml.suppl), which=1:4)
par(opar)
```

Observation 32 contains an extremely large value of `ml.suppl` and therefore has a large influence on its regression coefficient. Without `ml.suppl` in the model, the Cook's distances are much smaller.

12.9

```
juulyoung <- subset(juul, age < 25)
juulyoung <- transform(juulyoung,
                sex=factor(sex), tanner=factor(tanner))
fit.untf <- lm(igf1 ~ age * sex * tanner, data=juulyoung,
                na.action=na.exclude)
plot(fitted(fit.untf) ~ age, data=juulyoung,
     col=c("red","green")[sex])
fit.log <- update(fit.untf, log(igf1) ~ .)
fit.sqrt <- update(fit.untf, sqrt(igf1) ~ .)
opar <- par(mfrow=c(2,2))
plot(fit.untf, which=1:4)
plot(fit.log, which=1:4)
plot(fit.sqrt, which=1:4)
par(opar)
```

13.1

```
summary(glm(mal~age+log(ab), binomial, data=malaria))
```

(You may also want to check for interaction.)

13.2

```
attach(graft.vs.host)
type <- factor(type,labels=c("AML", "ALL", "CML"))
m1 <- glm(gvhd~rcpage+donage+type+preg+log(index), binomial)
m1a <- glm(gvhd~rcpage+donage+type+preg+index, binomial)
summary(m1)
summary(m1a)
```

The coefficient to `log(index)` is more significant, but the model with `index` has a slightly better deviance. There is little hard evidence for either. The log-transform has the advantage that it reduces the influence of two very large values of `index`.

```
drop1(m1, test="Chisq")
drop1(update(m1, ~ . - rcpage), test="Chisq")
drop1(update(m1, ~ . - rcpage - type), test="Chisq")
drop1(update(m1, ~ . - rcpage - type - preg), test="Chisq")
summary(m2 <- glm(gvhd~donage + log(index), binomial))
```

Notice that except for `log(index)` it is essentially an arbitrary decision which variables to put in the final model. Altman (1991) treats the `type` classification as separate binary variables and gets a final model where ALL and AML are combined into one group and includes `preg` but not `donage`.

13.3   For example,

```
confint(m2)
## normal approximation:
est <-  coefficients(summary(m2))[,1]
se <-  coefficients(summary(m2))[,2]
est + cbind(qnorm(.025)*se, qnorm(.975)*se)
confint.default(m2)
```

Notice that the `confint`-generated intervals lie asymmetrically around the estimate. In this case, both ends of the interval are shifted away from zero, in accordance with the fact that the deviance-based tests from `drop1` have lower *p*-values than the approximate *t* tests in `summary`.

13.4   The model can be fitted as follows

```
counts <- c(13,40,157,40,21,61)
total  <- c(108,264,375,310,181,162)
age <- gl(3,1,6)
type <- gl(2,3,6)
anova(glm(counts/total~age+type,weights=total, binomial),
      test="Chisq")
```

The effect of `type` vanished once `age` was included, suggesting that it really is the same disease, which has affected mostly younger (and fitter) subjects in the Eastern region.

13.5

```
juul.girl <- transform(subset(juul,age>8 & age<20 &
                              complete.cases(menarche)),
                    menarche=factor(menarche))
logit.menarche <- glm(menarche~age+I(age^2)+I(age^3),
                  binomial, data=juul.girl)
```

```
probit.menarche <- glm(menarche~age+I(age^2)+I(age^3),
                       binomial(probit), data=juul.girl)
summary(logit.menarche)
summary(probit.menarche)
Age=seq(8,20,.1)
newages <- data.frame(age=Age)
p.logit <- predict(logit.menarche,newdata=newages,type="resp")
p.probit <- predict(probit.menarche,newdata=newages,type="resp")
matplot(Age,cbind(p.probit,p.logit),type="l")
```

14.1

```
attach(graft.vs.host)
plot(survfit(Surv(time,dead)~gvhd))
survdiff(Surv(time,dead)~gvhd)
summary(coxph(Surv(time,dead) ~ gvhd)) # for comparison
summary(coxph(Surv(time,dead) ~
              gvhd + log(index) + donage + rcpage + preg))
```

Subsequent elimination suggests that `preg` might be a better predictor than gvhd.

14.2

```
attach(melanom)
cox1 <- coxph(Surv(days, status==1) ~
              log(thick) + sex + strata(ulc))
new <- data.frame(sex=2, thick=c(0.1, 0.2, 0.5))
svfit <- survfit(cox1,newdata=new)
plot(svfit[2], ylim=c(.985, 1))
```

14.3

```
summary(coxph(Surv(obsmonths, dead)~age+sex, data=stroke))
summary(coxph(Surv(obsmonths, dead)~sex, data=stroke))
with(stroke, tapply(age,sex,mean))
```

The men were considerably younger than the women when they had their stroke, which may explain their apparently better survival.

14.4   Using stroke2 from Exercise 10.4,

```
summary(coxph(Surv(entry, exit, dead)~age+sex, data=stroke2))
```

Notice that the result is essentially the same as in the unsplit analysis; only n and Rsquare are changed.

15.1   Using bcmort2 from Exercise 10.2,

```
bcfit <- glm(bc.deaths ~ (age + period + area)^2, poisson,
             offset=log(p.yr), data=bcmort2)
```

```
summary(bcfit)
drop1(bcfit, test="Chisq")
confint(bcfit, parm="period1981-1991:areaNat")
```

15.2 Continuing with `stroke2` from Exercise 10.4, the only slight complication is to convert `entry` to a factor to specify the relevant time interval.

```
summary(glm(dead~sex+age+factor(entry), poisson,
        offset=log(exit-entry), data=stroke2))
```

Notice how similar the results are to the Cox analysis in Exercise 14.3.

16.1 To fit to the data for girls, just copy the procedure for boys. Even though the growth curves differ, there is no real reason to redo the starting value calculation, so we can fit the model to boys, girls, and both as follows

```
girls <- subset(juul2, age<20 & age>5 & sex==2)
boys <- subset(juul2, age<20 & age>5 & sex==1)
young <- subset(juul2, age<20 & age>5)
stval <- c(alpha=exp(5.3),beta=exp(0.42),gamma=0.15)
fit.boys <- nls(height~alpha*exp(-beta*exp(-gamma*age)),
        start=stval, data=boys)
fit.girls <- nls(height~alpha*exp(-beta*exp(-gamma*age)),
        start=stval, data=girls)
fit.young <- nls(height~alpha*exp(-beta*exp(-gamma*age)),
        start=stval, data=young)
```

To test whether we can use the same model for boys and girls, there are two approaches. One is to make an F test based on the three fits above:

```
ms.pooled <- (deviance(fit.boys) + deviance(fit.girls))/(499+625)
ms.diff <- (deviance(fit.young) -
            deviance(fit.boys) - deviance(fit.girls))/3
ms.diff/ms.pooled
```

This gives $F = 90.58$ on 3 and 1124 degrees of freedom, which is of course highly significant.

Alternatively, we can set up the joint model with separate parameters for boys and girls and test whether it fits the data better than the model with the same parameters, like this:

```
fit.young2 <- nls(height~(alpha+da*(sex==1))*
                exp(-(beta+db*(sex==1))*
                exp(-(gamma+dg*(sex==1))*age)),
            start=c(alpha=exp(5.3),beta=exp(0.42),gamma=0.15,
            da=0, db=0, dg=0), data=young)
summary(fit.young2)
anova(fit.young, fit.young2)
```

Notice that da, db, and dg represent differences between the parameters for the two genders. The term sex==1 is 0 for girls and 1 for boys.

16.2 We consider experiment 1 only. Starting values can be eyeballed by using the observation at zero dose for y_{max} and the x (dose) value at approximately $y_{max}/2$ as β. The value of α can be guessed as the fixed constant 1. Below, we use α via its logarithm, called la.

```
e1 <- subset(philion, experiment==1)
fit <- nls(sqrt(response) ~ sqrt(ymax / (1 + (dose/ec50)^exp(la))),
       start=list(ymax=28, ec50=.3, la=0), data=e1,
       lower=c(.1,.0001,-Inf), algorithm="port")
summary(fit)
confint(fit)
p <- profile(fit, alphamax=.2)
par(mfrow=c(3,1))
plot(p)
confint(p)
```

16.3 The alternative model has similar tail behaviour but behaves differently when x is close to zero. (In particular, the original model has a term proportional to $-x^{\alpha-1}$ in its derivative. At zero, this is $-\infty$ when $\alpha < 1$ and 0 when $\alpha > 1$, so the model describes curves that are either very steep or very flat near zero.) Notice that, in the modified model, β is no longer the EC50; the latter is now the solution for x of $(1 + x/\beta)^{\alpha} = 2$. The two are connected by a factor of $2^{1/\alpha} - 1$.

```
e1 <- subset(philion, experiment==1)
fit1 <- nls(sqrt(response) ~ sqrt(ymax / (1 + dose/b)^exp(la)),
       start=list(ymax=28, b=.3, la=0), data=e1,
       lower=c(.1,.0001,-Inf), algorithm="port")
summary(fit1)
fit2 <- nls(sqrt(response) ~ sqrt(ymax / (1 +
       dose/(d50/(2^(1/exp(la))-1)))^exp(la)),
       start=list(ymax=28, d50=.3, la=0), data=e1,
       lower=c(.1,.0001,-Inf), algorithm="port")
summary(fit2)
```

Here, fit1 and fit2 are equivalent models, except that the latter is parameterized in terms of ec50. We can compare the fitted curve with the model from the previous exercise as follows:

```
dd <- seq(0,1,,200)
yy <- predict(fit, newdata=data.frame(dose=dd))
y1 <- predict(fit2, newdata=data.frame(dose=dd))
matplot(dd,cbind(yy,y1)^2, type="l")
```

(Notice that the fitted values should be squared because of the square-root transformation in the models.)

Bibliography

Agresti, A. (1990), *Categorical Data Analysis*, John Wiley & Sons, New York.

Altman, D. G. (1991), *Practical Statistics for Medical Research*, Chapman & Hall, London.

Andersen, P. K., Borgan, Ø., Gill, R. D., and Keiding, N. (1991), *Statistical Models Based on Counting Processes*, Springer-Verlag, New York.

Armitage, P. and Berry, G. (1994), *Statistical Methods in Medical Research*, 3rd ed., Blackwell, Oxford.

Bates, D. M. and Watts, D. G. (1988), *Nonlinear regression analysis and its applications*, John Wiley & Sons, New York.

Becker, R. A., Chambers, J. M., and Wilks, A. R. (1988), *The NEW S Language*, Chapman & Hall, London.

Breslow, N. E. and Day, N. (1987), *Statistical Methods in Cancer Research. Volume II: The Design and Analysis of Cohort Studies*, IARC Scientific Publications, Lyon.

Campbell, M. J. and Machin, D. (1993), *Medical Statistics. A Commonsense Approach*, 2nd ed., John Wiley & Sons, Chichester.

Chambers, J. M. and Hastie, T. J. (1992), *Statistical Models in S*, Chapman & Hall, London.

Clayton, D. and Hills, M. (1993), *Statistical Models in Epidemiology*, Oxford University Press, Oxford.

Cleveland, W. S. (1994), *The Elements of Graphing Data*, Hobart Press, New Jersey.

Cochran, W. G. and Cox, G. M. (1957), *Experimental Designs*, 2nd ed., John Wiley & Sons, New York.

Cox, D. R. (1970), *Analysis of Binary Data*, Chapman & Hall, London.

Cox, D. R. and Oakes, D. (1984), *Analysis of Survival Data*, Chapman & Hall, London.

Everitt, B. S. (1994), *A Handbook of Statistical Analyses Using S-PLUS*, Chapman & Hall, London.

Hájek, J., Šidák, Z., and Sen, P. K. (1999), *Theory of Rank Tests*, 2nd ed., Academic Press, San Diego.

Hald, A. (1952), *Statistical Theory with Engineering Applications*, John Wiley & Sons, New York.

Hosmer, D. W. and Lemeshow, S. (2000), *Applied Logistic Regression*, 2nd ed., John Wiley & Sons, New York.

Johnson, R. A. (1994), *Miller & Freund's Probability & Statistics for Engineers*, 5th ed., Prentice-Hall, Englewood Cliffs, NJ.

Kalbfleisch, J. D. and Prentice, R. L. (1980), *The Statistical Analysis of Failure Time Data*, John Wiley & Sons, New York.

Lehmann, E. L. (1975), *Nonparametrics, Statistical Methods Based on Ranks*, McGraw-Hill, New York.

Matthews, D. E. and Farewell, V. T. (1988), *Using and Understanding Medical Statistics*, 2nd ed., Karger, Basel.

McCullagh, P. and Nelder, J. A. (1989), *Generalized Linear Models*, 2nd ed., Chapman & Hall, London.

Murrell, P. (2005), *R Graphics*, Chapman & Hall/CRC, Boca Raton, Florida.

Siegel, S. (1956), *Nonparametric Statistics for the Behavioral Sciences*, McGraw-Hill International, Auckland.

Venables, W. N. and Ripley, B. D. (2000), *S Programming*, Springer-Verlag, New York.

Venables, W. N. and Ripley, B. D. (2002), *Modern Applied Statistics with S*, 4th ed., Springer-Verlag, New York.

Weisberg, S. (1985), *Applied Linear Regression*, 2nd ed., John Wiley & Sons, New York.

Zar, J. H. (1999), *Biostatistical Analysis*, Prentice Hall, Englewood Cliffs, NJ.

Index

Software for Data Analysis
Programming with R

John M. Chambers

This book guides the reader through programming with R, beginning with simple interactive use and progressing by gradual stages, starting with simple functions. More advanced programming techniques can be added as needed, allowing users to grow into software contributors, benefiting their careers and the community. R packages provide a powerful mechanism for contributions to be organized and communicated.

2008. Approx. 510 pp. (Statistics and Computing) Hardcover
ISBN 978-0-387-75935-7

Time Series Analysis
with Applications in R

Jonathan D. Cryer and Kung-Sik Chan

Time Series Analysis With Applications in R, Second Ed., presents an accessible approach to understanding time series models and their applications. Although the emphasis is on time domain ARIMA models and their analysis, the new edition devotes two chapters to the frequency domain and three to time series regression models, models for heteroscedasticty, and threshold models. All of the ideas and methods are illustrated with both real and simulated data sets. A unique feature of this edition is its integration with the R computing environment.

2008. 2nd Ed., 494 pp. (Springer Texts in Statistics) Hardcover
ISBN 0-387-75958-6

Data Manipulation with R

Phil Spector

This book presents a wide array of methods applicable for reading data into R, and efficiently manipulating that data. In addition to the built-in functions, a number of readily available packages from CRAN (the Comprehensive R Archive Network) are also covered. All of the methods presented take advantage of the core features of R: vectorization, efficient use of subscripting, and the proper use of the varied functions in R that are provided for common data management tasks.

2008. 164 pp. (Use R) Softcover
ISBN 978-0-387-74730-9

CPSIA information can be obtained
at www.ICGtesting.com
Printed in the USA
LVOW10s2118300817
546977LV00003B/120/P